DROUGHT-RESISTANT
PLANTING

LESSONS FROM BETH CHATTO'S
GRAVEL GARDEN

DROUGHT-RESISTANT
PLANTING

LESSONS FROM BETH CHATTO'S
GRAVEL GARDEN

BETH CHATTO

PHOTOS BY STEVEN WOOSTER

F

FRANCES
LINCOLN

To the memory of Andrew Chatto, whose lifelong study of the homes
of garden plants has inspired the making of the garden.

Frances Lincoln Limited
A subsidiary of Quarto Publishing Group UK
74–77 White Lion Street
London N1 9PF

Drought-Resistant Planting: Lessons from Beth Chatto's Gravel Garden
Previously published as *Beth Chatto's Gravel Garden: Drought-resistant planting through the year*
Copyright © Frances Lincoln 2000
Text copyright © Beth Chatto 2000
Photographs copyright © Steven Wooster 2000
except the following: p.13 © The Beth Chatto Gardens;
p.18 & 19 © David Ward; back cover (right) © Jerry Harpur

First Frances Lincoln edition: 2000
First Frances Lincoln paperback edition: 2016

A catalogue record for this book is available from the British Library

978-0-7112-3811-4

Printed and bound in China

9 8 7 6 5 4 3

Quarto is the authority on a wide range of topics.

Quarto educates, entertains and enriches the lives of
our readers – enthusiasts and lovers of hands-on living.

www.QuartoKnows.com

FRONT COVER *Nepeta*, *Allium* and *Bergenia* bloom in the Gravel Garden against a backdrop of ornamental grasses.

BACK COVER (Left) The proud stems of *Verbena bonariensis*. (Right) Beth Chatto among her plants.

HALF-TITLE Crowded heads of *Allium karataviense* crouch low above pairs of broad-bladed, purplish green leaves.

TITLE-PAGE After seven years' planting, the naked gravel flaunts a coat of many colours, textures and forms.

ABOVE LEFT TO RIGHT *Allium hollandicum* 'Purple Sensation', *Euphorbia myrsinites* and *Bergenia* 'Eric Smith'.

INTRODUCTION Silky tactile wands of the grass *Stipa tenuissima* provide a subtle contrast to the more vibrant
verticles of *Salvia nemorosa* subsp. *tesquicola* and the burgundy-velvet form of the campion, *Lychnis coronaria*.

Contents

INTRODUCTION TO THIS EDITION

Since its creation in 1990, Beth Chatto's Gravel Garden has surprised, inspired and educated us all. The early years of the experiment taught us quickly where the boundaries lay, which plants would survive and which would not. Despite our insistence on planting certain species that we believed would do well, we eventually learnt otherwise. We also came to understand that, as in their native habitats, many plants appear to enter survival mode in order to get through prolonged periods without rain. As a consequence, the Gravel Garden can appear (and indeed is) parched during drier summer months, with colours becoming muted when compared with the vibrancy of early summer when there is still moisture in the ground.

The phrase 'right plant, right place' is often used in association with the Beth Chatto Gardens and it is the backbone of our philosophy. Not only does it apply to the conditions a plant requires to reach its potential, but it is also appropriate when considering a plant's shape, size and aesthetic appeal in relation to its neighbours. For example, an Irish yew, *Taxus baccata* 'Fastigiata', planted in the Gravel Garden to form the high point of a small island bed was of course always going to grow higher, and it eventually outgrew the bed. Yes we could have regularly pruned it, but instead we sought more sustainable alternatives. Initially, we thought the adjacent giant feather grass, *Stipa gigantea* – a much more suitable height – would prove an effective substitute, but we realized any height gained in early summer would have been lost when its spent flower stems were eventually cut down in winter. Instead, we replaced the yew altogether with a sizable variegated *Yucca gloriosa* 'Variegata'. In our mind's eye it works, but only time will tell.

Other decisions are forced upon us. A freak gust of wind in the Scree Garden caused irreparable damage to the Judas tree, *Cercis siliquastrum*. It was a huge shame, but we have since learnt to seek new opportunities in seemingly adverse situations.

Indeed, one of gardening's great joys is seeking out new plants. Over the years, several additional species have come to establish themselves as key features of the Gravel Garden. For example, on rereading the chapter 'High Summer's Gauzy Veil', it is interesting to see that two of our now star performers during July were yet to make an appearance. Both the Californian tree poppy, *Romneya coulteri*, and the feather grass *Stipa barbata* have become such a staple in summer that it is hard to recall their absence. Other recent and invaluable introductions include the descriptively named *Sedum* 'Red Cauli' and the delightful new dark-leaved vervain, *Verbena officinalis* var. *grandiflora* 'Bampton'.

Beth Chatto's programme of annual renewal and the procedures practically and meticulously expressed in this book are still closely followed by head gardener Asa Gregers-Warg and her dedicated team. Beth's style of soil preparation, plant choice and meticulous aftercare routines have proven essential to rejuvenating and maintaining the vibrancy of this great garden.

Together we hope to continue the shared experience of both learning from and, most importantly, enjoying Beth Chatto's ongoing horticultural experiment in drought-resistant planting.

David Ward, October 2015.
Garden and Nursery Director
The Beth Chatto Gardens.

WHAT IS
A GRAVEL
GARDEN?

PREVIOUS SPREAD Pale pink
spikes of *Linaria purpurea* 'Canon
Went' usefully pop up through
the indispensably long flowering
spurge, *Euphorbia seguieriana*
whist in the background
Diascia integerrima and *Nepeta
racemosa* 'Walker's Low'
complete this summer scene.

How does a gravel garden differ from other forms of gardening? Since many people ask me this question, I want to begin by describing the prevailing conditions in my Essex garden because these govern the type of garden anyone can make, and by explaining the principles underlying my design and planting. These apply to the Gravel Garden, which I began in 1991, and to the more recent Scree Garden, where I have designed island beds to show off smaller plants, which tend to become overlaid in the main Gravel Garden.

Wherever we live, we all have to assess the amount or lack of water, the degree of heat and cold we receive. Whether or not you take global warming seriously, weather patterns *are* changing. We have kept records for forty years and our average rainfall is about 50cm/20in, roughly equal in winter and summer. In recent years we have had summers when the rain fell in dribbles, just enough to wash off the dust, during July and August. Despite occasional blips to the contrary, summers are hotter, with temperatures sometimes reaching 30°C/92°F. Winters are milder, with ice scarcely thick enough to bear a duck, in comparison with winters when weather forecasts reminded us to lay in stocks of parsnips and leeks before the ground froze too hard to dig them out.

Wherever we garden, we need to assess the soil. Ours is not derived from rock beneath but is a sedimentary soil, deposited by a melting glacier about 20,000 years ago. Great washes and rivers of melting ice must have gushed out over our east-coastal strip, piling up banks of gravel where the current was strongest and depositing fine sand, silt or clay where the water became shallow and still. In the Gravel and Scree Gardens, as on much of my 6.15 hectares/15¼ acres of land, the soil consists of gravel, a mixture of stones and yellow sand, looking just like the beach at nearby Frinton-on-Sea. Elsewhere in the garden we have silt and clay, which enables us to grow a different range of plants.

Given our prevailing conditions, a gravel garden seemed a good idea, but it is not a clever one of my own. Many people have made stone or gravel gardens, both before mine and since, and they have become fashionable in recent years. But my interpretation of a gravel garden is specific to my soil and conditions – it is *not* any piece of soil filled with an assortment of unrelated plants then covered with stones for effect. And although I grow a wide range of plants which can be grown in similar conditions around the temperate world, I must emphasize the word *similar*, since much higher temperatures or far lower winter temperatures than we have here would limit the range.

Visitors are always astonished when we show this barrowload of 'soil' in which the Gravel Garden plants are growing. It was removed from a trench dug through the garden when we laid all our overhead cables underground. The depth of gravel and sand is about 6m/20ft, overlying clay.

The point I need to stress is that copies of my Gravel Garden will not necessarily be successful or suitable if the principles underlying my planting designs are not understood. When visitors to my garden tell me they have attempted to make a gravel garden but the plants don't look or behave as they do in mine, they wonder what they have done wrong. I ask, What type of soil do you have? Very good, they reply. The amount of rainfall? Twice what we have here, they tell me. I laugh, and say if I had good loam and adequate rainfall I would not be growing drought-tolerant plants but would grow well-loved plants like delphinium, phlox and aster, and many more desirable species which thrive in such conditions. I have never gardened in an area where the average annual rainfall exceeds 63cm/25in, or on thin rocky soil on foothills or mountains. I am ill equipped to lecture gardeners living in wet Wales, on wind-shaven Scottish islands, in limestone villages of the Cotswolds, sun-scorched hillsides in California or maple-covered slopes in New England. But I imagine if I was transported there I would love to use the local stone, whether flat slates or limestone chippings, to protect the soil from the effects of leaching, from alternate wetting and drying, to inhibit the germinating of weed seeds, to conserve moisture and shade the roots of such plants, as would be suited to the conditions.

For some years I had dreamt of making my Gravel Garden, with plants adapted to the prevailing conditions, instead of watching mown grass turn biscuit-brown for weeks every summer. I hoped to see which plants would survive without hosepipe irrigation and was prepared to lose some of the many new introductions not yet sufficiently tested for summer drought or winter cold and damp. I would replace these with other, more resistant plants. I hoped to teach myself and possibly help visitors to make and maintain some kind of decorative garden without irrigation. I had in mind gardeners who might well feel aggrieved to see hosepipes lying about my garden, if they had been deprived of water for weeks.

So my Gravel Garden was to be a horticultural experiment. I am lucky to have room to experiment. Not all gardeners have such a wide canvas as I have here, but most gardens consist of sunny and shady areas requiring different treatments and plants suited to the conditions. Most people can deal with open areas and, by choosing plants growing in woodlands around the world, even dark areas need not be rank and featureless simply because roses and dahlias will not thrive in them. Shady places may well become a favourite part of the garden, as they have in mine. Success depends on some knowledge of plant provenance and on an understanding of natural plant associations.

During the last fifty years, there has been emerging, like a tidal wave, a surge of interest in species plants – plants as they are found growing in the

OPPOSITE The Gravel Garden is so named because the depth of virgin gravel and sand is about 6m/20ft, overlying clay. Such a hungry soil cries out to be fed by the addition of organic matter to give young plants the best possible chance to establish.

wild. Before the Second World War, cultivars were the proper plants for good gardens, introduced by horticulturists, who 'improved' wild plants until they were larger, brighter, more double than anything seen before. Some undoubtedly were improvements, and still are desirable. Vast numbers of species plants have been, and are still being, introduced. Not all are garden worthy. These have transformed our approach to garden design and are an important part of our palette, but often on receiving a new plant we find we know little or nothing of its needs, and nothing at all about the plants it was found growing with. Was it sheltered in bushy scrub, in deep black meadow soil or on a sun-baked bank? This information would help all gardeners in their choice of appropriate plants.

My understanding of plant associations adapted to prevailing conditions has come from my husband Andrew's life-long study of the origins of garden plants, and from holidays abroad where I have been thrilled to see plants I recognized from my garden thriving in their native habitat or environment.

Andrew's interest in the ecology of garden plants began more than seventy-five years ago when he lived, for a few years, as a young boy on the west coast of California. For a while he and his parents lived at Laguna Beach, then a tiny fishing village, but over the past fifty years it has developed into a wealthy residential area, where, in recent times, bush fires have ravaged the homes of famous film stars. Andrew wandered through the untouched hills inland and saw plants that were familiar to him because he had seen them growing in gardens in England: blue ceanothus, azalea, artemisia and golden carpets of eschscholzia, the orange and yellow Californian poppies. How did they get to California? The reply, of course, was that these plants were native there, as elderberry and blackberry bushes are in England. Although he became a fruit-grower by profession, this experience sowed the seed of Andrew's interest in the ecology of garden plants – that is their natural associations in the wild. This involves not only the country the plants come from, but also the conditions: the climate, aspect, soil and all the plants which together create an association for that particular species.

Of course, in gardening we cannot reproduce Nature and, on the whole, it would not be desirable. Few people find beds of nettles or thickets of blackthorn attractive, even though they have their place in Nature's scheme of things and must be conserved where possible. We all find a haze of bluebells beneath beeches, primroses on clay soil beneath oaks or a damp meadow golden with buttercups more magical than anything we can create. However, in our gardens we look for more. We learn to make plant associations that extend the season, to create pictures worth living with throughout the year.

Soon after the Second World War we had our first holiday abroad, to Saas Fee in the Valais, Switzerland. In those far-off days we walked with rucksacks on our backs from the valley road below to reach this mountain village, now transformed by tourism. We were obliged to lean into the rocky mountainside to make way for strings of mules, loaded with panniers of goods, picking their way up and down the rock-strewn path. It was in the High Alps above the village – too high, we fervently hope, to be spoilt by modern conveniences made for leisure-seekers – that I was introduced to Nature's natural gardens, just below the permanent snow-line. I saw for the first time that, whatever the site, whether stony slopes exposed and windswept above high mountain passes, north-facing slopes or damp hollows and streamsides, each had its own distinctive group of plants, adapted to differing situations. Each group made pictures I have never forgotten. Baking screes and moraines, a-flutter with life and colour in May or June, were being watered from underground, by unseen snow-melt from glaciers above, so the crowns of many plants remained dry while they had adequate moisture at the root during the growing season. Many plants I recognized and could understand at last why we were not growing them successfully: plants such as trollius, persicaria and thalictrum. We were far too dry. And our rock garden lacked conditions needed to retain the characteristics of true high alpines – tight hummocks pressed into crevices studded with tiny jewel-like flowers. When transported to kinder lowlands these plants, no longer disciplined by harsh conditions, deteriorate into untidy straggling mats, totally out of character.

In the autumn of 1989 Christopher Lloyd and I were invited to give a series of talks that would involve us in a month's journey around the world. Each country we visited merited at least a month to itself, to begin to observe and absorb the variety of life, both plant and human, that unfolded around us. We had to make do with fleeting glimpses, a few hours, a day or two only, spent here and there. But the memories remain, haunting me.

In New Zealand we found the foothills of the Southern Alps were curiously covered with low tussock-like plants creating a monotonous peppercorn effect with bare brown earth showing between, the same colour as many of the grass-like plants, reminding me of pictures of a bushman's head when the bare scalp shows between tight knots of black curls. As we climbed higher more varied plants appeared, including small compact bushes with leathery leaves. Recognizing the sharp green tones of our first hebe was the moment to leave the car and explore on foot. Below a rocky ledge almost the first plant I thought I could identify appeared to be *Cassinia leptophylla* subsp. *fulvida*, whose tiny leaves and stems gilded with yellow felt form graceful evergreen bushes 90–120cm/3–4ft high. Nearby was something very like our familiar blue-leafed *Hebe pinguifolia* 'Pagei', while providing a contrast in green were spreading shrublets of a gaultheria which might not be hardy with us, but the North American *Gaultheria procumbens* could be substituted, and all these plants could make attractive contrast and companions to heathers in any but the coldest of gardens, provided they were grown in well-drained soil. Tucked around the shrubby plants in gritty soil were carpets of acaena and silvery trails of several different raoulias, while to my delight I found a colony of the tiny bronze-leafed *Geranium sessiliflorum* subsp. *novae-zelandiae* 'Nigricans' seeded about in open spaces. As is often the case in mountainous regions, plants which might appear baked and sun-scorched are often cooled and moistened by upward currents of warm air from the lowlands meeting cold air in the uplands, consequently releasing moisture. We had first-hand experience of this as clouds descended, drenching us in chilling mist.

Later we had a picnic lunch sitting on piles of water-rounded boulders, in a partially dried-up river bed where the idea first came to me for the design of my Gravel Garden-to-be. Bothered by sand flies, we wandered out on to an open stony beach and there found a colony of *Leptospermum scoparium* hanging over a bank, whose small aromatic leaves, we were told, were used by early settlers to make tea. In its cultivated form it is grown in sheltered British gardens, presenting an irresistible temptation to buy when seen at the Chelsea Flower Show with flowers single or double, white, red or pink clustered tightly among small leaves on slender drooping branches. Related to leptospermum is *Melaleuca alternifolia* (both are members of the family Myrtaceae), which is the source of tea tree oil, popular for its antiseptic and healing properties. Along the roadsides, we saw groups of *Griselinia littoralis* with its apple-green, soap-smooth leaves, and *Pittosporum* (pronounced in New Zealand as Pitt<u>os</u>porum, with the emphasis on the '*os*'). Both these evergreens can be grown in British gardens, but are liable to be lost in very cold winters.

Among other plants we saw were phormium, brachyglottis, olearia and aciphyllas, with both narrow and sword-like, silver-grey leaves. Most covetable were the celmisias, which vary tremendously. Some have large handsome rosettes of grey, tongue-shaped leaves while others embroider stony slopes with little rosettes of leaves smothered in white daisy flowers. (It is a curious fact that most flowers in New Zealand are white.) I am tempted to add celmisias to my list of plants in the scree beds, but so far experience tells me I should not risk killing them by too dry conditions. Otherwise, most of these plants I saw growing in the wild in New Zealand are now flourishing in either the Gravel Garden or the Scree Garden.

Several years after beginning my Gravel Garden I was staying with Christopher Lloyd, and, together with young friends, we took ourselves off for the day to Dungeness on the Kent coast. Inevitably we botanized, noses to the ground as we wandered around, marvelling at such plants as could survive the seemingly hostile conditions: wave-battered stones and incessant wind. More beautiful, indeed spectacular, than I have ever seen before were colonies of sea kale, *Crambe maritima*, their stout branching flower stems supporting large bouquets of pea-sized seed cases. Most attractive, they never set seed like that in my garden! Christopher was also tickled by my chagrin at finding foxgloves and wood sage, *Teucrium scorodonia*, flourishing where, to my mind, they had no business to be, since normally they are found on the edges of woodland. The only 'woodland' visible on this vast beach of cobblestones being, of all things, isolated groups of stunted hollies carved into bizarre shapes by the wind. As we walked to and fro, stumbling over the stones, I suddenly found myself looking down on to an unexpected woolly ball of *Santolina chamaecyparissus* growing straight out of the beach, followed by other grey-leafed plants, and poppies and pinks, all mixed up with curious artefacts made from flotsam and jetsam blown in by the winds and tide. Looking up I realized we were in someone's garden, although there had been no barrier to cross, only near by a small fisherman's cottage, its wooden walls painted black, the door and window frames canary-yellow. Christopher, enthralled and oblivious, was busy with his camera. Another moment's pause, and I was mortified. We were not unobserved. A young man appeared, listened to stammered apologies, then disappeared. Almost immediately a tall lean figure, not a fisherman, appeared. It was the film producer Derek Jarman, with a warm welcome for two people he had never met before, instant joy in sharing with us his passion and delight in finding plants which could survive on raw shingle, perhaps 50m/163ft from the shoreline. I heartily regret I had no notebook to record the variety of plants which appeared to thrive in such an unexpected terrain, but the memory of meeting a unique and courageous man has never left me. I felt he was sustained by the miracle of growing plants.

Having seen plants growing in such a hostile environment, I came away from Dungeness with renewed determination to go on with my experiment. There will be disappointments and failures, but every day I feel a sense of wonder and delight at what plants can do if given a chance. We hope to go on experimenting and extending our knowledge of plants, to be able to grow them to their advantage and for our peace of mind.

BEGINNING THE GRAVEL GARDEN

I have been growing drought-resistant plants in my present garden for almost forty years, and I am thankful there are so many lovely plants which survive, last well and reliably clothe my garden in spite of regular periods of drought. But in the autumn of 1991, I faced a bigger-than-usual decision concerning their use. We were removing our visitors' cars from the grass car park which had lain in front of our house for twenty-five years to a newly prepared one where I had bought an extra piece of land.

What should I do with the approximately 0.35 hectares/¾ acre of rather tatty grass, heavily compacted by cars? As we have possibly the lowest rainfall in England, and the climate appears to be becoming even drier, it seemed senseless to prepare the land for more grass, even though it was a temptation to do so because running a mower over would, initially, seem to be less trouble than a new garden. But I knew the soil was poor and to prove it we dug an inspection trench, about 45cm/18in deep, and there, below the top few inches of dark soil, we found unbelievable yellow sand and gravel. This layer continues down approximately 6m/20ft deep until it meets the underlying strata of clay. This base, combined with low rainfall and desiccating wind, means that grass inevitably burns to toast in summer, no matter how much compost is incorporated. In place of grass I would choose to grow decorative plants adapted to these poor conditions.

We began in the autumn of 1991 by killing the car-worn grass with glyphosate. After several weeks, when the turf had changed to brown, I laid out the design for my Gravel Garden on to this blank canvas, using hosepipes to define the borders and island beds.

I do not care to plan on paper, but prefer to work on site, shaping outlines with all available lengths of hosepipe stretched along the ground. The area, roughly rectangular, had no straight edges so a formal design seemed inappropriate. I had in mind a dried-up river bed as I made a gently sinuous walkway through the length of my piece of land, breaking up the long curving borders thus planned with several spur exits and entrances, and leaving space in the central walkway for large island beds for the gravel to swirl around. I found I needed to outline several island beds at the same time, and to consider the relationship of one bed to another, while also looking ahead to the next section to make sure the central pathway and subsidiary paths flowed unobtrusively. Without this attention to the overall appearance, the effect might have been a series of odd beds scattered about, rather than a harmonious whole. I made a rough sketch of the whole area with important measurements written over it, then all the hosepipes were removed while the soil was prepared.

In other parts of the garden, we hand-dig or trench beds to make renovations, but here digging was not practical on such a large scale. First we used a subsoiler, to burst up the compacted soil. This splendid tool, sharply angled like the back leg of an insect, can sink 60cm/24in into the soil, breaking up the hard pan to allow penetration of rain and roots. Next the whole area was ploughed to turn in the grass sods. A few weeks later, in January 1992, we borrowed a farm roller to lightly flatten the furrows so I could lay out my initial design on a flat surface using all my accumulated hosepipes.

Knowing how dry this soil could become, we thickly covered the surface of the beds with home-made compost, mushroom compost and bonfire waste. This would help conserve moisture and would be incorporated to a depth of two spits to provide nourishment in the first vital months. Compost incorporated two spits deep might seem excessive for drought-tolerant plants, but we are not gardening on average garden soil. It is more like gardening on the beach. It is unfair to expect plants to put up with the worst you can offer and we wanted to give them a good start. Once established, the roots would be well down, where less rich conditions would promote tough and wiry growth, making plants more resistant to both drought and winter cold.

The photograph taken in January 1992 shows how the compacted soil has been broken up, the design outlined with hosepipes and the beds thickly spread with compost.

PLANTING IN THE FIRST SEASON

By the beginning of March 1992, the soil had settled, the paths were gravelled and my blank canvas was ready for planting. I had a large plastic tank of water centrally placed where I could stand my trolleyloads of plants, dip every pot and watch bubbles of air rise until the rootballs were saturated. It is wise to do this with any plant you buy from a garden centre or nursery. First dunk it in a bucket till the bubbles stop. Although it may look damp the pot of compost can be so dry it will float before absorbing enough water to fill up the air spaces and swell the dried-out compost. Each plant was well watered in, using a hose for the first and last time.

I did not apply a mulch while I was planting in the first year, since I know by experience that fluffed-up soil and incorporated compost encourages a healthy crop of weed seedlings. I find this problem best solved by hoeing the first season. Once a mulch is laid it should be disturbed as little as possible. The beds were hoed with swoes, in my experience the best kind of long-handled hoes to use since they cut through the soil with an easy push-and-pull action. To work carefully round small plants, I prefer a short-handled onion hoe, and one which has been worn down by several years of hoeing, rather than the broad clumsy things we buy as new! During the first exciting months, my women gardeners kept the swoes going before weeds could think of setting seed, enjoying as much as I the rapid development of the planting.

Back to March and my empty canvas. At this stage, I feel very much as I do when I take a sheet of blank paper and wonder what words and sentences I shall choose to write. The analogy can be extended into the garden. I take a story, say the Dry Garden, and then proceed to use groups of plants (phrases or sentences) which look well together and are adapted to the conditions, and then punctuate them with feature plants, making emphasis.

In some situations, such as public parks and large gardens, I think it is best to use fewer species and to plant in bold groups so as to avoid the moth-eaten carpet effect which can result when unsuitable plants die out. However, in my own garden I confess to being a collector. I enjoy creating a rich tapestry effect for as much of the year as possible, using the soil vertically as well as horizontally, so plants are coming up throughout the growing season from different depths, as earlier ones, having completed their cycle, are dying down, making way for the next phase. This applies particularly to bulbs, but to some herbaceous plants as well. My approach to planting the Gravel Garden was to achieve these layers of interest over time, allowing seasonal effects to build up as my experiment progressed.

In any planting scheme I like to begin with the dominant plants around which to base the rest of the design. The area was already enclosed by the wall-like clipped hedge of Leyland cypress, × *Cuprocyparis* leylandii (3m/10ft high), which protected it from searing winter winds from the east, while a well-matured curving border of trees and shrubs framed the western side. A few, not too many, well-placed but light trees were needed to make accents in this seemingly vast flat plain. To stand out against the hedge and provide a living look in winter I planted several eucalyptus trees, *E. gunnii* and *E. dalrympleana*. They might become unwieldy but could be pruned to form handsome background shrubs. (The first year they shot up and blew over, so in spring 1993 I cut them off about 1m/3ft high to make them break again and form more stable bushy shapes. We have not kept up the pruning; they now are securely rooted and stand well above the hedge, breaking the flat plane formed by the clipped top.)

As exclamation marks I planted *Juniperus scopulorum* 'Skyrocket ' and the bulkier *Cupressus arizonica* 'Pyramidalis'. I planted several specimens of the Mount Etna broom, *Genista aetnensis*. Elsewhere in the garden this has formed a graceful shape more than 6m/20ft tall, dripping in July with showers of tiny, sweetly scented, yellow pea flowers.

Among feature-forming shrubs I chose the bladder senna, *Colutea orientalis*, with divided, blue-green leaves and clusters of small, brick-red pea flowers followed by pale green, inflated seed pods. Once it was established, I would plant one of the forms of a Viticella clematis to scramble through it.

From a friend's garden I was given an unnamed form of the smoke bush, *Cotinus coggygria*, which, once it gets its roots well down, would stand the droughts, since the type can be found rooted in rock fissures hanging over the Italian lakes. This plant is possibly a selected seedling, not a very dark purple-leafed form, rather a softer, more gentle shade of purple.

For general furnishing, I selected plants which will survive in free-draining soils with sparse summer rain, primarily those found growing wild in countries bordering the Mediterranean. They include old friends like lavenders in various forms, attractive white-felted *Ballota pseudodictamnus*, and the grey, woolly-leafed *Santolina chamaecyparissus*. I also used *S. rosmarinifolia* for contrast among the grey and silver plants since it makes patches of vivid green which can come as a relief among too much ashen-grey.

Where the borders are deep, bold groups of these plants reached almost to the back, where eventually they would be smothered by larger permanent shrubs like *Berberis* × *ottawensis* f. *purpurea* 'Superba', arbutus, large cistus and forms of lavatera, all chosen for their tolerance of dry conditions. Artemisias in all shapes and sizes are useful fillers among heavier neighbours, from the large feathery shrubs of *A. arborescens* to low silky mats sprawling across the pathways. *Brachyglottis* (Dunedin Group) 'Sunshine' (syn. *Senecio* 'Sunshine'), from South Island, New Zealand, eventually forms bold bushes of oval, almost white-felted leaves, and is invaluable as an accent among smaller fussier plants. To grow up through it I planted *Clematis* × *durandii* to provide a perfect setting for the deep violet-blue flowers. Another clematis, *C.* 'Étoile Violette', scrambles about on the ground among the white daisies of *Anthemis punctata* subsp. *cupaniana*, and twining into the hazy spires of Russian sage, *Perovskia* 'Blue Spire'. It has encouraged me to plant more small-flowered clematis to provide colour in late summer (see pages 93–4).

With ample humus clematis do well once their roots are well down but I doubt if they would thrive in poor soil in hotter climates. Our summers are often dry, sometimes reaching 30°C/92°F but rarely for long periods.

When making new borders one always needs disposable fillers to cover bare spaces and check weeds, while sprawlers are essential to soften hard edges. I welcomed some of the sprawlers to trail into the gravel paths. Various forms of thyme are ideal used in this way. Rock roses, helianthemums, in shades of pink, red, yellow and white, make low, wide-spreading bushes, their small evergreen

or greyish leaves contrasting with the silky ears of that old cottage garden plant lambs' ears (*Stachys byzantina*). In just a few months *Gypsophila* 'Rosenschleier' (syn. *G.* 'Rosy Veil') formed a billowing froth of tiny, double pink flowers, placed beside the serpentine-like trails of *Euphorbia myrsinites*.

Too many rounded shapes herded together can be monotonous. To avoid this I look for good verticals. They may be boldly dramatic like *Verbascum bombyciferum* with large felted leaves and huge candelabra-like heads, studded with small soft yellow flowers. Or they can form delicate screens through which to view something on the other side of the border. The *Perovskia* 'Blue Spire' does just that; its many upright delicate branches carry close-set fuzzy buds that open into small lavender-blue flowers and create a soft haze of colour for weeks in late summer. *Verbena bonariensis* takes the same shade into autumn; its bare angled stems forming a delicate lattice-like effect topped with flat heads of tiny vivid mauve flowers set in dark calyces.

I also use bulbous plants to rise as verticals above the basic furnishing plants. Various alliums include *Allium hollandicum* and the extraordinary wild leek, *A. ampeloprasum*, which grows taller than any of the other ornamental onions. I placed groups of these far back in the borders between ballotas and cistus, where their cricket-ball-sized heads of silvery pink flowers would show against the dark green leylandii hedge. *Nectaroscordum siculum* subsp. *bulgaricum*, *Gladiolus communis* subsp. *byzantinus* and *Fritillaria persica* would all emerge as star performers in early summer, their fading foliage masked by santolina, artemisia or succulent clumps of sedum. There would be lilies too, after the oriental poppies had flowered, while the green- and white-flowered galtonias, *Galtonia viridiflora* and *G. candicans*, would add fresh elegance in late summer.

Not all my colour schemes are soft and muted. From midsummer onwards the most vibrant groups are accented by kniphofias. Over many years I have selected good forms from seedlings, so now we have a wide range of sizes and colours, from rich reds, through flaming orange, soft apricot to cool lemon, and even green.

Big bold accents are needed to create 'punctuation' among so many fussy small-leafed plants. The first season I cheated by bedding out several large pot-grown forms of *Agave americana*, both grey-leafed and variegated forms. They looked so good that I was almost tempted to plant them out annually, but for dramatic effect even in winter I planted hardy yuccas instead. Now, in 1999, the yuccas make impressive features throughout the year. And the agaves? They are kept in frost-free shelter during winter and return to the garden as living sculpture, making bright accents throughout inevitable weeks of summer drought.

At the end of the first summer in 1992, I wrote: 'We are all surprised and delighted at how well everything has grown. The hot dry spell in May and June was a testing time, but welcome rain came in time while it was still warm for everything to put on new growth so that already this new garden has an established look. During the summer I used several half-hardy plants like green-flowered tobacco, *Nicotiana* 'Lime Green', to help fill in gaps, while the prettiest little yellow daisy-like plant, *Bidens ferulifolia*, sprawled like sunshine across the gravel paths until frost extinguished it. There will be much to learn from this experiment. Not all plants will be successful, some may die, others may prove unsuitable, or simply it may be I won't like the effect, or the way one thing smothers out another. Making a garden is like working on a canvas, painting in highlights and shadows, never can one say it is finished.'

MULCHING AND MAINTENANCE

In the second spring, 1993, while there was still some moisture in the soil, a mulch of 12mm/½in gravel (brought in from a local quarry) was spread between the plants, 25–50mm/l–2in deep, according to the type of plant. Very low-growing plants cannot be buried – and in any case most of them colonize the ground quickly, making a natural mulch. I admit I was concerned about adding a lot of stone to my already gravelly soil, so beneath shrub and tree plantings, in the back or middle of borders where the effect would be less noticeable, we used baled straw, brought in from a neighbouring farm and put down in late autumn. It was teased out and then thickly spread like an eiderdown, to prevent weeds germinating, help preserve moisture and, as it broke down, add a little nutrient. Of course it was most conspicuous when the trees and shrubs were small, and some observers commented unfavourably on the appearance; but my first concern is always for the welfare of my plants. I knew that in a few months the farmyard effect would diminish as the fresh straw turned brown with decay, and it delighted me when exposed soil elsewhere was dust-dry to find the soil beneath the straw was cool and damp beneath its protective cover.

In subsequent years we have organized ourselves better so that we have weathered straw bales, partially rotted, which are less conspicuous, although heavier to handle. We renew the eiderdown when it becomes tatty, which might be after more than one year, depending on the degree of decomposition. This treatment might not be necessary in reasonably fertile soil – it must be remembered that in our free-draining gravel nutrients are quickly absorbed or washed down into the 6m/20ft depth of gravel and sand which are the strata beneath us.

The summers of 1995 to 1997 brought abnormally high temperatures, accompanied by merest dribbles of rain. During those periods the soil beneath the mulches dried out. My resolve not to irrigate almost deserted me one afternoon when hot winds had blown from the Sahara for several days and the plants were almost brittle with stress. Although I was wilting visibly myself, David Ward, my nursery manager, urged me not to lose heart, saying we must continue the experiment else all we had done thus far would be wasted. I knew he was right. Within a short while the wind swung round and cooler nights brought dew, which gave some relief, as did lower temperatures. The plants responded and within a few days of the first penetrating rain (which usually comes on the August Bank Holiday) every plant refurbished itself with fresh young growth. The crisis was past, the effect miraculous.

I do not think our Gravel Garden is labour intensive in comparison with some other types of garden, particularly other large spaces, such as parks, where substantial areas are bedded out, and where the hoeing of weeds day after day, month after month, must be as continuous and boring as painting the Forth Bridge. In the first few years, before the plants had covered the ground, there was weeding to be done, but not as much as if there had been no gravel mulch. Now after seven years much of the original bare 0.35 hectares/¾ acre is covered with plants we have chosen to grow. Many shrubby plants, such as cistus, santolina, ballota, helichrysum and helianthemum, together with many smaller mat-forming plants, not forgetting wide sweeps of bergenia for contrast, have made very efficient ground cover.

The close planting, combined with gravel mulch, means there is relatively little weeding of ordinary garden weeds, yet inevitably the odd sow thistle will occasionally rise up in front of me flaunting its yellow flowers, to assert its right to be there, I suppose. Of course, weeds can always be found! Most persistent and tiresome is the ubiquitous 'spit-in-your-eye' cress, whose minute white flowers produce in no time at all long thin seed pods all set to shoot ripe seed deep into the crowns of emerging perennials, or peppered over mats of carpeting plants. It is all too easy to miss the next generation of trouble makers. We do what we can, but will never eradicate them.

Other weeds are mostly our chosen plants – euphorbias, arums, opium poppies, love-in-a-mist, alliums – all lovely things which can unbalance the design, smother other plants if too many are allowed to stay in the 'wrong' place. It is not always easy to make the right decision, whether to leave enough to create a natural effect, or to leave too many and create chaos. But self-sown seedlings of *Euphorbia characias* subsp. *wulfenii* must be spotted and removed if they are likely to smother something else I value. There are places here where self-sown euphorbias have created a picture I had not dreamed of, but they need to be controlled, not allowed to take over. Other self-seeders include the wild poppy *Papaver rhoeas* Mother of Pearl Group and the equally desirable umbellifer *Bupleurum falcatum*. On the whole they are not troublesome. The effect between more stable neighbours is so attractive when such plants, mostly annuals, occur in natural drifts.

We simply do not have time in the normal running of the gardens and nursery to keep records of time spent on general maintenance. But since I was concerned about the extra work involved in maintaining an additional area of complex planting, I asked for records to be kept. Here they are:

1994	Cutting down, grooming and weeding	111 hours
	Ongoing weeding and spreading gravel mulch	177 hours
	Spreading straw	7 hours
1995	From late winter to July	
	Grooming and weeding	139 hours

I wonder how these figures would compare with a similar area planted with bedding plants and bulbs such as in public parks.

We have not kept records subsequently. As plants have grown, more time is taken to prune, shape and groom (see pages 35 and 160) and to plant replacements where needed, and less time is required for weeding. Such maintenance requires thoughtful and observant gardeners, but not constant attention. On average the Gravel Garden is attended five or six times a year; it takes about a week each visit to do all that is necessary to preserve a natural yet cared-for look. Covering the soil with plant growth (and mulches) for most of the year creates a picture in every season – and does most of the weeding for us.

Overall Plan   Island bed no.3

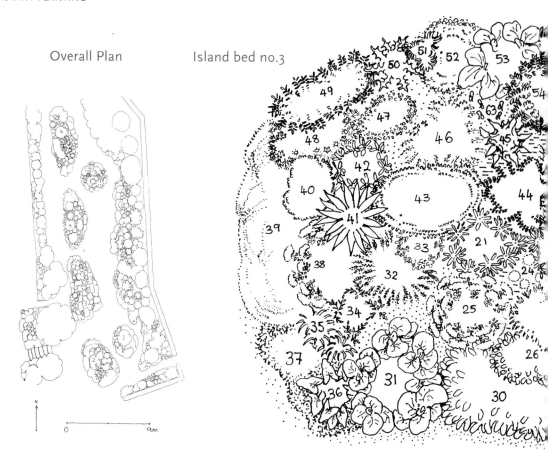

GRAVEL GARDEN PLANS

The overall plan represents an informal planting of 0.35 hectares/¾ acre, which has matured into a landscape of heights and valleys. Deep borders on either side merge visually with six island beds, and gravel paths give continuity throughout the area.

I have chosen to provide a detailed plan for island bed No. 3, since the planting fits comfortably together and is effective in summer and winter. It is a balanced design, using bulbs, perennials and shrubs – thus involving shape, texture and colour, of which enough is retained all the year round to maintain a harmonious whole.

The plants are listed in groups to show which look well together and could be chosen to create pictures in smaller areas.

Group One

1   *Berberis × ottawensis* f. *purpurea* 'Superba'
2   *Euonymus fortunei* 'Emerald 'n' Gold'
3   *Amelanchier lamarckii*
4   *Phlomis tuberosa* 'Amazone'
5   *Calamagrostis × acutiflora* 'Karl Foerster'
6   *Sedum* 'Matrona'
7   *Stipa calamagrostis*
8   *Bergenia* 'Admiral'
9   *Gaura lindheimeri* 'Siskiyou Pink'
10 *Artemisia absinthium* 'Lambrook Mist'
11 *Perovskia* 'Blue Spire'
12 *Agapanthus campanulatus* 'Cobalt Blue'
13 *Euphorbia epithymoides* 'Major'
14 *Origanum* 'Norton Gold'
15 *Allium schoenoprasum* f. *albiflorum*
16 *Sedum populifolium*
17 *Diascia barberae* 'Blackthorn Apricot'
18 *Helianthemum* 'Wisley Primrose'

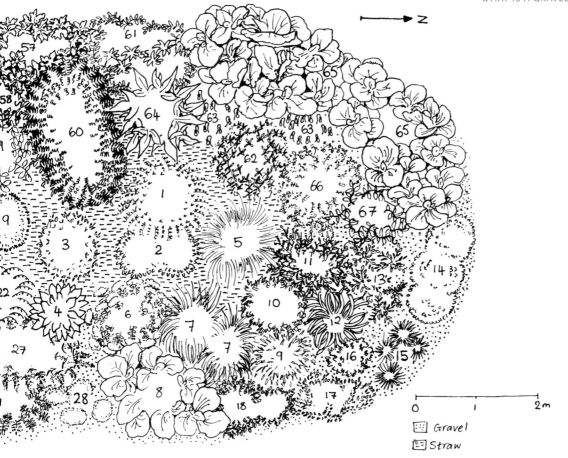

☐ Gravel
☐ Straw

Group Two

19 *Juniperus scopulorum* 'Skyrocket'
20 *Clematis* 'Madame Julia Correvon'
21 *Euphorbia characias* subsp. *wulfenii*
22 *Cytisus* × *praecox* 'Albus'
23 *Caryopteris* × *clandonensis* 'Worcester Gold'
24 *Allium hollandicum* 'Purple Sensation'
25 *Sedum* (Herbstfreude Group) 'Herbstfreude'
26 *Cistus* × *pulverulentus*
27 *Nepeta racemosa* 'Walker's Low'
28 *Thymus* 'Silver Queen'
29 *Helianthemum* 'Wisley Primrose'
30 *Stachys byzantina* 'Primrose Heron'
31 *Bergenia* 'Morgenröte'

Group Three

32 *Lavandula angustifolia* 'Twickel Purple'
33 *Lespedeza bicolor*
34 *Euphorbia epithymoides*
35 *Sternbergia lutea*
36 *Arum creticum*
37 *Phyla nodiflora*
38 *Sedum spectabile* 'Iceberg'
39 *Phuopsis stylosa*
40 *Helichrysum hypoleucum*
41 *Yucca gloriosa*
42 *Alstroemeria ligtu* hybrids
43 *Hebe* 'Red Edge'
44 *Cotoneaster franchetii*
45 *Helleborus argutifolius*
46 *Spiraea japonica* 'Bullata'
47 *Codonopsis clematidea*
48 *Nigella damascena* 'Miss Jekyll Alba'
49 *Helianthemum* 'Rhodanthe Carneum'
50 *Acanthus dioscoridis* var. *perringii*
51 *Ipheion uniflorum*
52 *Origanum vulgare* 'Aureum'

Group Four

53 *Bergenia* 'Abendglut'
54 *Hebe* 'Nicola's Blush'
55 *Colchicum* 'Rosy Dawn'
56 *Vinca minor* 'La Grave'
57 *Geranium macrorrhizum* 'Ingwersen's Variety'
58 *Clematis* × *triternata* 'Rubromarginata'
59 *Lonicera* × *purpusii*
60 *Hebe* 'Nicola's Blush'
61 *Vinca minor* 'Atropurpurea'
62 *Berberis julianae*
63 *Galanthus* 'Washfield Warham'
64 *Helleborus* × *sternii* 'Boughton Beauty'
65 *Bergenia* 'Mrs Crawford'
66 *Rosa spinosissima* 'Falkland'
67 *Clematis* 'Étoile Violette'

THE FRESH COLOURS OF SPRING

In March, if I miss being in the garden for a couple of days, I feel as if it has been a month, since so much change happens overnight. In early spring, the heavenly blues of chionodoxas, scillas and irrepressible ipheions are at their best. The blue-and-white striped *Puschkinia scilloides* var. *libanotica* from the Caucasus, just opening, looks well against the pale stones of the gravel. Young beetroot-red shoots of *Euphorbia dulcis* 'Chameleon' contrast darkly with the deep blue of the scillas. From among the tufts of grassy leaves the pale washed-blue stars of *Ipheion uniflorum* will peep for weeks to come. New velvet-soft rosettes of lambs' ears (*Stachys byzantina* and *S. b.* 'Primrose Heron') are pushing through the shabby remains of last year's leaves, fritillaries rising out of their young growth, and alliums are springing up between succulent rosettes of sedum. *Nectaroscordum* is 30cm/12in high already and in the blink of an eye this strange allium will have flower buds enveloped in papery, pointed envelopes looking like candle-snuffers.

All over the area, from front to back of beds, bulbs are bursting through. They have so much to do, to produce good foliage, to make and nourish new bulbs while there is ample moisture and sunlight. Many, including the alliums, will have done their work, will be collapsed and dried before the flower stems appear. Camassias make lovely verticals among grey ballotas and santolinas, while Madonna lilies, *Lilium candidum*, rise nearby to take their place. I brought seed of *Camassia leichtlinii* subsp. *suksdorfii* Caerulea Group – the blue form, from Portland, Oregon, where it grows wild. It opens earlier than the cream-coloured form.

By the end of March and throughout April the Gravel Garden looks amazingly lush, full of new growth – an incredible appearance of fertility in spite of poor soil and low rainfall. I am delighted how well the snowy mespilus, *Amelanchier lamarckii*, has done on this site. The abundant trusses of small white, cherry-like flowers are warmed by the young foliage, which is bronze-tinted as it opens. *Hippocrepis emerus* (syn. *Coronilla emerus*), although backed by the leylandii hedge where it must become dry in summer, is nearly 2m/7ft tall and 3m/10ft across, crowded with tiny yellow pea flowers smothering the finely divided foliage, all borne on slender branchlets.

Always outstanding at this time of year are the euphorbias, their electrifying greens making an exciting combination with bergenias on either side of my river of gravel. Both will play an important role for months to come, the euphorbias holding centre stage well into June and major contributors to the garden's lush freshness. By then the wild tulips that set the curving border edges on fire will have come and gone. Love-in-a-mist, *Nigella damascena*, seeded around from last year's flowers, pushes persistently through the collapsing foliage of early spring bulbs, while bold-leafed plants like the shield-shaped leaves of *Arum creticum* contrast with mounds of feathery foliage. Soon the area will be highlighted with the beautiful lemon-yellow spathes of this arum, which has survived several years in the warm, well-drained gravel soil. Repeating this, yellow, tight bunched heads of *Euphorbia epithymoides* gradually lengthen and expand till they completely cover the fading leaves of bulbs.

WAKING UP WITH A START

You can almost see the lush foliage of alliums developing as you watch. Most of them have long blade-shaped grey-green leaves, tucked behind or between small bushes or emerging herbaceous plants where their dying remains will not detract from flower heads in April and May. But an unusual allium has caught my eye just pushing through the stones. It is *Allium karataviense*. From each bulb, two broad-bladed leaves appear, clasped together, grey-green above, deep purple below. By late April they will have extended to form a cradle for the short-stemmed, pinky beige flower heads, in size varying from a golf ball to an orange. But, in early spring, these bold-shaped leaves make much-needed contrast where they have seeded among a variety of small-leafed plants: *Origanum* 'Kent Beauty', the woolly-leafed *Thymus doerfleri* and *Geranium subcaulescens* 'Splendens' with tiny parasol-shaped leaves. The overwintered strap-shaped leaves of *Sternbergia lutea*, dying down now, are reminders of the yellow crocus-like flowers to come in late autumn.

By the last week in March the young leaves of *Lonicera × purpusii* are overwhelming the remains of sweet-scented cream flowers. At its feet, still in flower and scattered throughout the area (see the plan on pages 28–29), is a rather special snowdrop. It makes hefty clumps of wide grey-green leaves, with well-shaped single broad-petalled bells hanging well above the foliage. It is very free flowering, giving at least a month's display until the end of March. It has now been identified as *Galanthus plicatus* 'Washfield Warham'. The robust and squeaky-fresh bunches of colchicum leaves are mingled among the clumps of snowdrop, reminders of the purple goblet-shaped autumn crocus which will take their place in September.

Nearby is a huge plant of *Helleborus × sternii* 'Boughton Beauty'. This plant is a hybrid combining the greater size and hardiness of *H. argutifolius* with the amazing plum shades of *H. lividus*. It is now about six years old and I have just counted over fifty flower heads. The maroon-stained stems radiate out like the spokes of a wheel, each stem weighted by a large cluster of cup-shaped flowers, shaded maroon and apple-green, filled with cream stamens. As they gradually open new buds over many weeks, the heads expand to display more and more

TOP LEFT *Helleborus × sternii* 'Boughton Beauty' was named after Boughton House, home of Lady Scott, better known as Valerie Finnis, gifted gardener and plant photographer. Photographic geniuses cannot do justice to this strangely beautiful plant as, over a period of many weeks, the flowers become suffused with shades of cinnamon-pink.

flowers, creating a vast garland of flowers around the perimeter of the 'wheel'. This, to some minds, untidy habit of growth makes way for the purple-stained young shoots emerging from the central crown. My flower clusters rest their chins on the sturdy clusters of colchicum leaves or nestle in carpets of periwinkles. The handsome foliage of the hellebore will remain reasonably upright until late autumn, leaving open spaces around it for the colchicum flowers to emerge in September.

Making good contrast with the lush foliage of alliums are feathery mounds of *Ferula communis*. Although sometimes called giant fennel, this differs considerably from its relative the common fennel, *Foeniculum vulgare*. Its dark glossy green leaves are not so finely shredded, nor are they at all aromatic. Instead of making a crowd of flower stems, it takes several years developing a crown and rootstock sufficient to raise one great stem, covered in purplish bloom, fading to green as it matures. Rising high above everything else, this noble architectural feature needs to be planted as a youngster, about the thickness of a pencil, the spot marked with a cane to guard it – especially when dormant, when you might want to plant something else in the 'empty space' – until the time, two to three years further on, when you will be astonished by the flowering plant.

A plant that confounds me now in the Gravel Garden is *Arum italicum* subsp. *italicum* 'Marmoratum'. In my young gardening days it was considered something of a rarity, needing, so I was told, a cool leaf-mould soil in shady conditions. Walking along the base of the leylandii hedge, behind the groups of cistus, abutilon, berberis and lavatera established there, I am astonished by colonies of this beautiful arum growing luxuriously on the very roots of the shrubs, with huge leaves 30cm/12in long and across at the base. The leaves are dark green and glossy, vividly marbled in an exquisite pattern. The main veins form meandering rivers of pale green, while tributaries of little veins add a finer network of pale silvery green over the dark green base, and a plain narrow border trims the wavy edge of the leaf. I did not plant arums here, and can only think they have been brought in when compost has been spread, or have been dropped by birds. Both I imagine.

A friend coming to the garden on Good Friday for bits and pieces for the church flowers found these unusual leaves the perfect complement for the arum lilies she was using for the Easter Sunday service. The degree of variegation varies considerably in this arum, as does the size of the leaf. Throughout winter I can find tiny leaves ideal to put with snowdrops, aconites and sweet-scented twigs of *Lonicera* × *purpusii*, and now, in late March, crowded colonies growing in grass still produce small leaves. By May, after flowering, the leaves collapse, to reappear with the rains in September.

As children my brothers and I looked for sweet violets, both purple and white, on ditch banks and along hedge bottoms. Planted here in part shade and leafy soil the flowers are lost in a wealth of foliage, while on some of my most exposed gravel they form purple carpets of blossom before the leaves have had time to develop. Like the arum, violets too upend my view of what is best for them.

PRUNING AND GROOMING

March is the time to prune and groom overwintered foliage before new growth becomes too advanced. I still find it hard to cut down the bleached grasses which have given me such pleasure throughout the winter, but when the word has been given, and the deed done, I am relieved, and within a few days am amazed to see several centimetres of fresh young shoots emerging from the basal clumps.

The aim with most shrub pruning is to encourage and maintain a strong vigorous framework. I reduce this theory for most shrubs to three positional cuts. The first is low down, to remove dead or worn-out wood near the base, and any weak or twiggy growth. This will let in light to encourage strong replacement growth from the base. The second aim is to thin out some of the middle of the bush, again removing weak or crossing branches. Lastly I often tip the tallest shoots, either to control size or strengthen the previous year's leaders, which may eventually become part of the renewed framework. This simple method ensures light and air throughout the bush and, since whenever you make a cut you encourage new growth, the bush is regularly furnished with new shoots. At the same time enough of the previous year's wood will be left to maintain a shape and give good flower and foliage – and maybe provide a support for a scrambling plant such as a small-flowered clematis.

Berberis × *ottawensis* f. *purpurea* 'Superba' responds to thinning and reducing overall size with brighter, larger foliage. Once every second year is enough. The new growth of *Brachyglottis* (Dunedin Group) 'Sunshine' is so attractive it would pay to tip this several times a year to keep it both shapely and full of fresh silver-grey leaves. But if your bush is ragged and woody, you can prune to a framework in spring. Then in mid- to late summer tip the shoots to encourage new growth, which will mature before winter.

Caryopteris × *clandonensis* is pruned now to form a basic framework about 45cm/18in tall, since it will have all summer to make shapely bushes smothered in blue fluffy flowers among grey-green leaves by late summer.

Cistus can become leggy and untidy, and smother other things if ignored too long. Shortening back to a lower shoot, thinning and cutting out worn or dead pieces after flowering all helps. We had a huge bush of *Cistus* × *pulverulentus* which had become bare and woody in the centre, spreading too far. Should we hard prune or remove it? We pruned hard and were surprised and delighted by the new growth coming from bare, woody stems.

The silver-leafed shrub from the foothills of the Adas mountains, *Atriplex halimus,* needs reducing in early spring, sometimes by as much as a third since it grows vigorously and can take more than its share of space. Carefully done with secateurs so that it does not end up looking like a huge dumpling, it will continue to make a graceful feature until it needs restraining again, in August.

The soft tip shoots of ballota were damaged by the odd frost, and although they look untidy I am reminded to wait another week or two before we remove these straggly pieces, which could help protect the flush of new velvety leaves already forming beneath them.

Hippocrepis emerus (syn. *Coronilla emerus*) can make a huge bush where suited, but needs pruning after spring flowering every few years even where there is space enough, to keep a strong, vigorous framework. Remove old wood from the base, reduce some of the stems and cut back remaining branches by a third, removing weak or crossing stems. By September you will be surprised and delighted to find the bush alive again with little pea flowers produced on the new summer growth.

Russian sage, *Perovskia* 'Blue Spire', too has its slender whitened stems cut back almost to ground level where already new growth shoots are breaking.

The leafless shrubs of *Prunus triloba* are showing tight incurved flower buds which will open pale rose-pink double cherry blossoms well into the middle of April. We grow this plant from cuttings rather than grafting it, so there is no problem with tiresome suckering shoots. After flowering it is hard pruned (even though it will already be making fresh growth) to encourage the production of long uninterrupted wands of blossom next year. If not pruned the existing stems produce many side shoots, resulting in a muddled twiggy effect which is not so attractive.

Ruta graveolens needs to be pruned in spring to encourage plenty of fresh growth. I once left some bushes so bare after taking cuttings they looked liked Brussels sprout stalks, and was amazed at the subsequent outburst of fresh growth. But do wear gloves, and keep your sleeves rolled down. Rue is very toxic, and can cause an irritating rash and blisters.

All forms of *Salvia officinalis* tend to make lax, woody stems, holding clusters of new shoots. Frequent light pruning would keep them ideally clothed with fresh growth, but we never have time for that. They survive for years with minimum attention. Sometimes we remove raggedy pieces, but eventually make the decision to replace them with young specimens.

FASCINATING FRITILLARIES

One of the first signs of spring, early in March, is a sudden strange musky scent caught unexpectedly on the cold air as I set off on some errand round the garden. It comes from the great crown imperial fritillary, *Fritillaria imperialis,* even before the fat buds have penetrated the soil. Some people say it is unpleasant, and that it smells of foxes. Although I have seen foxes on several occasions in the garden, I cannot say I have smelt them. I like this strange pungent smell. Every year it makes my pulse quicken as I recognize the first sign of feverish activity going on just beneath my feet. Within a few weeks there will be more to see, smell and touch than we can possibly keep up with. These isolated happenings can be savoured before the flood overwhelms us. Provided March remains free of hard frosts and snow, these dramatic plants will be in flower before the end of the month. They increase well in my stony soil, forming generous clumps. Each stout leafy stem is topped with a bunch of pointed, strap-shaped leaves (somewhat like a pineapple tuft), beneath which hangs a coronet of bell-shaped flowers in shades of burnt orange or lemon-yellow. If you tip up the downturned bell you will see a large drop of nectar at the base of each petal. Tradition has it that this haughty flower refused to bow its head at the Crucifixion. It has bowed and wept ever since.

Another early fritillary, *F. verticillata,* found wild in Central Asia and China, makes a delicate feature growing through carpets of yellow-leafed lambs' ears, *Stachys byzantina* 'Primrose Heron'. It is also tucked in around the wandering shrublets of dwarf Russian almond, *Prunus tenella,* soon to be crowded with bright salmon-pink flowers, which have *Arabis alpina* subsp. *caucasica* 'Flore Pleno' trailing among them to cover the bare lower half of the stems. This fritillary's slender stems (50cm/20in tall) carry spires of pale green, downturned shallow bells, netted green on the outside, delicately chequered inside, with touches of green, brown and soft maroon. At the base, the stems are clothed in narrow, pointed, grey-green leaves, becoming narrower as they ascend the stem, till at the top, among the bells, they emerge as curled crooks. For what purpose were these designed, I wonder? Was it to enable the flowers to be held well above the surrounding herbage? What do these bulbs grow among in the wild – in grassy places, or scrubby bushland? Some forms of this fritillary are shy flowering but we have a form, originally from my old friend and mentor, the late Sir Cedric Morris, which freely produces stems of flower 50cm/20in tall.

Fritillaria tuntasia is a native of Greece, found in rocky places and scrubland. Each stem, 15–20cm/6–8in long, carries several open bell-shaped flowers, smooth and sloe-black on the outside, slightly corrugated inside. In my garden, it has interbred, possibly with *F. messanensis,* which has larger and longer bells of a warm chestnut-brown colour, so we have variation in flower size and shape.

Another fritillary, *F. pyrenaica*, compels you to bend down to study its quaint beauty. On slender stems 40cm/16in tall, narrow, grey, lance-shaped leaves diminish in size to the top where hangs down a large downturned flower, deep brownish purple with a dusky bloom, shaped like a pull-on cap with rolled brim where the joined petals curl back to show a shiny yellow lining. Tip up the flower and you will see the enamelled surface lined with faint green stripes disappearing into a dark heart, darkly chequered purple. Protected in this mysterious area are stigma and stamens loaded with creamy pollen. A little taller than *F. pyrenaica* is *F. acmopetala* with green and chestnut-brown, bell-shaped flowers held on slender stems, sparsely clothed in narrow grey-green leaves. It is found wild in Eastern Europe in scrub and cornfields.

One of the most exotic-looking bulbs as March turns to April is *F. persica* 'Adiyaman'. Planted here about four years ago, each bulb has divided, so forming good groups of flowering stems standing 75cm–1m/1½–3½ft, clothed in their lower half with pointed narrow grey leaves, whipped by the west wind into curving shapes. Above them the upper half of the bare stem carries a spire of thirty to forty shallow bells, dark as purple grapes. They are perfectly supported behind by a big sprawl of purple-leafed sage, *Salvia officinalis* 'Purpurascens', which repeats their dark tones. In another group they are growing through *Alstroemeria ligtu* hybrids, whose young growth already helps to support them, as do the succulent clumps of *Sedum* 'Matrona' in the foreground. Its purple-stained leaves will mark where the fritillary bulbs lie resting till next spring. Behind this group a huge shrub, *Berberis* × *ottawensis* f. *purpurea* 'Superba', is laden with buds about to open, while various catmints, lambs' ears and verbascums are hurriedly crowding into the scene. There will be no gaps when the fritillaries have died down.

WILD TULIPS AND APRIL SHOWERS

On the whole I prefer to use species bulbs (bulbs that have occurred naturally in the wild), as opposed to cultivars which have been developed by horticulturists. In a different setting, such as in a public park, large, bright-coloured tulips are needed and look right. And so they do in many garden settings. I particularly like the striped and feathered tulips, but keep them for the early spring–summer pot gardens. As yet I haven't felt easy about letting them into the Gravel Garden. The following species thrive in my gravel soil, increasing every year, mostly placed among low carpeting plants along the curving border edges.

Tulipa dasystemon increases well to form clumps of long narrow leaves. The flowers, several to a stem, have creamy petals, rusty-backed, which open to the sun like pointed stars, followed by handsome seed pods. They stand about 15cm/6in tall above carpeting plants.

Tossing in the wind are the delicate lily-shaped flowers of *T. orphanidea* with long, pointed petals suffused with tones of cream and burnt orange, deeper toned inside where pollen-laden stamens nestle in brown-shadowed hearts. They stand 25–30cm/10–12in tall, coming through a carpet of *Leptinella potentillina*, its flat-pressed ferny leaves dotted now with minute puffs of pale green flowers.

Tulipa hageri has globular-shaped flowers on short 22cm/9in stems, the colour warm cardinal-red softened by olive shading outside, brown shadows within.

Tulipa linifolia has narrow grey leaves with slightly puckered edges; they lie, like star-fish, almost flat against the soil as if to capture every ray of sun to feed the bulbs, which make narrow-petalled scarlet flowers opening flat in early spring sunshine, showing dark stamens against a bruised purple centre.

Tulipa saxatilis runs about making underground stolons which form new bulbs. Lilac-pink in bud, it opens flat to the sun to show a rich yellow heart which sets off black anthers surrounding a green ovary. This tulip needs full sun and poor stony soil to have a good baking; otherwise in richer conditions you will have many leaves but few flowers. A sun-baked spot crowded with short-stemmed flowers opened to the sun is a sight to treasure.

TOP *Tulipa dasystemon*

BOTTOM *Tulipa orphanidea*

TOP *Euphorbia epithymoides*

BOTTOM *Euphorbia characias* subsp. *wulfenii*

OPPOSITE It is exciting to have both the space and conditions to show off the variation in form of *Euphorbia characias* subsp. *wulfenii*. Seedlings are rarely identical. Flower heads may be large and rounded or narrow and cylindrical, all in varying shades of green.

ELECTRIFYING EUPHORBIAS

In comparison with many other flowering plants whose effect is so ephemeral, there are few which give such value over long periods as euphorbias. We grow many members of the family in the garden here, some of which need retentive or even boggy soil, while others relish some shade in leafy soil. These would wither and die in my Gravel Garden conditions, so are not included here.

In the Gravel Garden the excitement begins early, in late winter, as *Euphorbia characias* subsp. *wulfenii* slowly unrolls its bowed and 'feathered' necks. Gradually the tightly clustered buds open to form huge cylindrical heads of shallow-cupped flowers of the brightest yellowy green. It is the showiest plant in the garden for weeks on end, literally for months, until June, continuing to make a dramatic feature while spring flowers come and go. When they are spotlit from the east in the early morning, and again in the evening from the west, I stand lost in wonder as these noble plants seem to reflect light into my kitchen, especially seen against a dark thunderous sky. As foliage plants alone they make architectural features for the rest of the year. Sometimes I drive past a small front garden where one of these plants grows as the sole feature, an architectural contrast to neighbouring borders packed with pansies and wallflowers. I am lucky to have the space to use them in drifts, often self-seeded. From seed you will find considerable variation in colour tone. Some can be a rather leaden green, probably with a dark or even black eye. *Euphorbia characias* itself, from Eastern Europe, is a rather sinister-looking type, with narrow heads of closely packed flowers. Most forms of *E. c.* subsp. *wulfenii* available as garden plants have some of this blood in them, which can produce interesting variations. But for garden effect the brightest yellow-toned forms are very popular. Named forms, such as *E.c.* subsp. *wulfenii* 'John Tomlinson' (collected in Greece), which shows colour earliest, are grown from cuttings, and consequently are not so readily available as those grown from seed. Over the years we have isolated good bright forms and plants grown from their seed come relatively true. The variation occurs mainly in the size and shape of flower head. Good plants can carry twenty to thirty flower stems or more, standing 1–1.5m/3½–5ft tall, with flower heads 30cm/12in long and 15–22cm/6–9in across. They make the substance and impact of a medium-sized shrub. We have an extraordinary variegated form of this euphorbia called *E.c.* subsp. *wulfenii* 'Emmer Green'. The whole plant – leaves, stems and flowers – appears cream from a distance. A closer look reveals leaves heavily margined and veined with cream; the flower cups too are almost devoid of green. Such strong variegation reduces size and growth, making the plant suitable in a place near the front of the border. The past few comparatively mild winters have not tested this euphorbia for hardiness.

The good drainage in the Gravel Garden helps *E.c.* subsp. *wulfenii* to withstand cold winters, but over many years we have learnt to expect some losses, especially among old specimens when hard frost has persisted for several days. Seedlings survive unscathed. But gardeners with heavy soil and inclement weather must expect losses. It is found wild in the southern Mediterranean, from Portugal and Morocco eastwards to south Turkey.

Another form is *E. characias* 'Portuguese Velvet'. Instead of waxen bloom-coated leaves, every part – stems, leaves and flowers – is covered in short pile (almost invisible downy hairs), velvety to touch and producing overall a faintly glistening effect over the dark blue-green foliage. Long, narrow, cylindrical heads of flower are the same dark tone accented with almost black centres, reminding me of the words of E. A. Bowles (quoted by Graham Stuart Thomas in his invaluable *Modern Florilegium* apropos *E. characias*): 'Curious dull heads of flowers with their conspicuous black spots . . . while I like to call it the name I learnt (in Dublin), the Frog Spawn Bush.' I think that almost certainly *E. characias* 'Portuguese Velvet' needs a well-drained soil with its back to a sun-baked wall to survive a serious winter. Concerning the 'dull effect' of the *characias*-type euphorbias, I was interested to see recently how an imaginative flower arranger had combined both bright and 'dull' forms in an all-green arrangement in our village church. As with all plants it is a matter of placing them, to make the most of individual characteristics.

However, there are plenty of smaller euphorbias to delight, which do come through the winter successfully. The sinuous trails of *E. myrsinites* clothed in grey-blue scale-like leaves are topped in winter with close rosettes of flower buds. For weeks they expand and open lime-green flowers eventually smothering their leaves with wide-spreading flower heads, making wonderful contrast for scarlet *Anemone × fulgens*, seeded among them or the little wild tulip, *Tulipa linifolia*, which is the same intense colour; but the anemones last longest until they are transformed into spools of cotton wool full of ripening seed and their place is taken by the vertical stems of *Anthericum liliago* 'Major'. This group is backed and enhanced by the large-leafed form of *Salvia officinalis*, while the large leaves of a bergenia form a support and define the path edge.

Nearby is a similar, but seldom seen euphorbia, *E. rigida*, more upright than *E. myrsinites*, with whorls of pointed, scale-like leaves along its rose-tinted stems. Its flowers are a richer, more golden green, and as they mature the saucer-shaped bracts turn warm shades of coral-red as the seeds they carry develop. Although the plant itself appears to be quite hardy in our well-drained soil, the likelihood of it setting good seed is threatened by frost, since the flowers appear so early. We cover it with a polythene cage when night frost is forecast. Later the fat seed pods, as big as green peas, will be protected in paper bags, to prevent the pods suddenly exploding. The native home of this plant is south-east Europe. It must be a wonderful sight to see a colony growing on rocky limestone, on shale and schist slopes.

Euphorbia nicaeensis from the southern Mediterranean looks somewhat between *E. myrsinites* and *E. rigida*. More upright than *E. myrsinites*, its red stems, bare at the base, clothed above with narrow grey leaves, carry very rounded heads of jade-green flowers later than the first two. It prolongs the season with its spring-like colour, but needs shelter from severe winter cold. It is a very lovely plant, but suitable only for favoured gardens.

While *E. dulcis* is a fairly dull little thing until autumn, when its foliage suddenly takes on fiery tints, *E. dulcis* 'Chameleon' is remarkable from the minute it pushes through the soil low mounds of bronzy, beetroot-red leaves. Seedlings come true. Their intense dark colour, scattered haphazardly, creates a patchwork-quilt effect among bright, open feathery stands of love-in-a-mist, succulent pewter leaves of *Sedum telephium* subsp. *ruprechtii* and low cushions of *Saponaria ocymoides* already embroidered with tiny pink campion flowers. Purple-stained calyces shelter minute green-tasselled 'eyes', while autumn sees an intensification of this dark colouring into a blaze of cherry and plum shades. (However, I have now been forced to accept that the 'soil' in my Gravel Garden is too poor for this euphorbia. In a weakened state, it became disfigured with rust so has been moved elsewhere – but seedlings continue to appear.)

Euphorbia epithymoides and *E. e.* 'Major', as eye-catching as daffodils, are similar, *E. e.* 'Major' being somewhat larger. Both make lowish tufts of stems, forming border-edge bouquets of flattish heads, crowded with bright yellow bracts and flowers. Another variation, *E. e.* 'Candy', has young foliage tinted bronze. In the dry conditions of the Gravel Garden they are sometimes disfigured later in the season with mildew. When we remember, we cut them to the ground after collecting the seed, and they quickly refurnish themselves, to our relief, since the new growth restores the ground cover. Both these euphorbias die down to basal buds in winter, so are not affected by cold.

Euphorbia pithyusa makes a low woody framework of stems and branchlets, covered in whorls of small, pointed, grey-blue leaves. I like it for its foliage effect, both shape and colour, with low tulips, like the subtle brownish red of *Tulipa hageri*, threaded through it. Later, its wide heads of small green flowers are not particularly exciting, but can create another texture above flat mat-forming plants such as thymes or *Silene uniflora*, a form of the sea campion.

Euphorbia seguieriana subsp. *niciciana* is a blessing because it does not begin to flower until the spring and early summer euphorbias are spent. Much more

OPPOSITE Low evening light casts a pattern of sunshine and shade across this spring-time scene. The rose-pink flowers of *Bergenia* 'Rosi Klose' add warmth to the cool greens of various euphorbias, while the tall columns of leylandii cypress make the necessary architectural support.

delicate-looking, with many slender stems forming a free and graceful group, it gradually opens out into a huge bouquet as the flat lace-like heads of small pea-green flowers develop, retaining their fresh colour well into autumn. As their leaves tire and drop, red-tinted stems add another texture rising out of the gravel floor. An annual euphorbia, *E. stricta*, contributes waves of spring-like green in high summer (see page 91).

Euphorbia cyparissias is feared because of its invasiveness. Fair enough in small areas, but where there is room, in semi-wild places, especially in a rock crevice, its dainty habit of growth and acid-yellow flowers are delightful. I have seen a high mountain slope in Portugal crowded with *Narcissus poeticus*, with this euphorbia (or something very similar) like a waving sea beneath them. Another form (still a colonizer) called *Euphorbia cyparissias* 'Fens Ruby' enchants me with its foliage effect, perhaps like small conifer seedlings, or bottlebrushes. Its many stems are clothed on dark blue-grey leaves, the top tufts stained maroon, a nice contrast for the lacy-looking heads of yellowy green flowers. Along our entrance drive, leading to the house, it can do little harm with such competing neighbours as bergenias and periwinkle, in particular the lovely early-flowering *Vinca minor* 'La Grave'.

It is important to remember the white milky sap in euphorbias can easily spurt or dribble from freshly cut stems. It is damaging to skin, especially on sensitive areas. Wear gloves, and be careful of your eyes. When cutting for a flower arrangement, to prevent the cut ends drying and to sear the stems, stand them, after recutting, in hot water, or hold them over a gas flame for a few moments. They then last well in water.

Most euphorbias set good seed. When the seed capsules are ripe you can hear them exploding, making little popping noises on warm afternoons. Extra special ones we enclose in paper bags until they have ripened. Some selected and named forms which would not come true from seed are propagated by cuttings.

Euphorbias are practically trouble-free, provided you choose those best suited to your conditions, and, unless you have ideal soil, you prepare the site adequately by deep digging (to break the pan beneath) and adding a little humus (or gravel in stodgy clay) to give them a good start. Once they are established you will be rewarded with self-sown seedlings flourishing in places you would not have dreamt of putting them, such as tucked into a crack between paving and the base of a wall. Perversely such plants often do better than any we have carefully provided for, but, in general, I find plants are often healthier and stronger when and where they have chosen to put themselves.

LEFT Among the almost black introductions of border iris, this elegant beauty *Iris* 'Black Swan' holds upright silk standards and drooping black velvety falls on 75cm/ 30in stems.

RIGHT *Iris* 'Pearly Dawn' is shorter, carrying more branched heads of translucent chiffon-textured petals in creamy pink tones, deepening towards the centres, making me think of a fruit sorbet.

ADDITIONAL TOUCHES

After writing descriptions for our catalogue of some good introductions of border iris, flourishing in the nursery stock beds, I wanted to try them in the Gravel Garden. I hoped they would not look too exotic among some of my species plants. I find I have to be careful how I mix cultivars and species plants together without detriment to one or the other. By early summer, surrounded by a wealth of shapes and textures, the irises might add just the right touch of ephemeral glamour, as do the good German hybrids of oriental poppy – and indeed the single and double forms of opium poppy, *Papaver somniferum*, self-sown and suddenly exploding among mounds of cistus, phlomis, salvia and ballota.

Included among the new irises we planted in spring 1998 is *Iris* 'Black Swan', with black velvet falls and upstanding purple, silk-textured standards. It is backed by bushes of lavender, the succulent *Sedum* (Herbstfreude Group) 'Herbstfreude' and *Euphorbia epithymoides*, which will carry rusty-red seed pods. Another iris, 'Jane Phillips', makes strong-growing clumps, very free-flowering, with ruffled petals of sky-blue. Behind it – and later – will be a blue sea holly, *Eryngium* × *tripartitum*, and the tall yellow hollyhock, *Alcea rugosa*, with *Euphorbia myrsinites* in the foreground, interplanted with the delightful annual *Omphalodes linifolia*, whose tufts of narrow grey leaves carry showers of small round white flowers.

Lastly we planted *Iris* 'Pearly Dawn', a shade of pink to die for, for flower arrangers I'm sure. Short stems (not too short to look stunted) carry a succession of pale, translucent pink flowers, the petals reminding me of a fruit water ice with a hint of apricot. Already the sword-shaped leaves make strong accents among mound-shaped plants, such as santolinas and bush sage, or with lower plants like the grey-felted leaves of the orange Californian daisy, *Eriophyllum lanatum*.

Towards the back of the wide border backed by the leylandii hedge I tucked more infant plants of the gigantic fennel, *Ferula communis*. As I did so I nodded appreciatively to the plants David had put in last year. Looking like teenagers waiting to take the place of mature specimens, they are now large foaming masses of finely cut leaves preparing to flower next year. Each will send up stems over 3m/10ft tall. (Many years ago we cut down such a stem, I lay next to it on the path and it measured more than twice my height.)

We planted three pot-grown specimens of *Astelia chathamica* (syn. *A. c.* 'Silver Spear'), their stiff blade-like leaves washed with silver patina, to make dramatic accents among dark, leathery cistus leaves and a brooding mounded form of ceanothus, all backed by the silver variegated privet *Ligustrum ovalifolium* 'Argenteum'. We must wait to see how the astelias survive winter conditions, but I am hopeful for them in this well-drained soil, with shelter from the hedge.

Thinking of the weeks and months to come I could see we needed a few more edging plants. For some reason not all the rock roses, Helianthemum, had survived. Perhaps they were sitting in a pocket of pure gravel. We pushed aside the gravel mulch, dug out some of the stony soil, added a good dressing of well-rotted compost and replanted with proven varieties, such as 'Rhodanthe Carneum' (syn. 'Wisley Pink') and 'Sudbury Gem'. We added white and purple forms of *Lychnis coronaria* to make vertical shapes above mats of *Phuopsis stylosa* and behind *Helianthemum* 'The Bride' to repeat her white flowers on another level – probably after the shell-pink oriental poppy, *Papaver orientale* 'Karine', planted alongside, has faded.

SUMMER'S ABUNDANCE

PREVIOUS SPREAD Although
I grow predominantly species
plants I enjoy planting cultivars
among them where they can
make impact without looking
too out of place. In Germany,
on the nursery of my late friend
Helen von Stein Zeppelin, her
head gardener Isbert Preussler
worked for many years producing
new forms of oriental poppy.
I particularly value those with
medium-sized flowers held
on stems needing little or no
support. This picture shows
Papaver orientale 'Juliane'.

Unlike a painting, a garden constantly changes, is never the same, from one day, even from one hour, to the next, especially in early summer when multitudes of plants are thrusting towards the light, determined to do their duty and spread their genes.

It is so beautiful and so ephemeral, from early morning to the last rays of sun falling in strips and shadows. As I write this, on a warm day in late May, our resident pair of cuckoos have been flying and calling back and forth, the collared dove's soft mournful tone contrasting with the busy positive tweeting of many other small birds. Thrushes and blackbirds fly in and out of every bush, busy feeding their young. A pair of fly-catchers skims my head as I sit having tea beneath the magnolia, their nest very close to that of a thrush, as they both rush in together with beaks full. I love the fly-catchers' dipping flight, rising and falling over the Gravel Garden. Their curiously long tails and manner of flight, so easily recognizable, reminds me of little clockwork toys we had as children, which bobbed up and down before the mechanism was broken.

We have not had a measurable drop of rain this month but several periods of unusually high temperatures and too much wind. However, the entire garden including the Gravel Garden, local gardens and the countryside around can never have looked more beautiful, with such an abundance of blossom, from trees, shrubs and flowering plants, or so lush, the good growth generated by steady, gentle rain in April (100mm/4in), spread over three weeks, so it could penetrate and soak down to the roots. (Why do the weathermen speak of *threat* of rain?)

I stand often and marvel at the exquisite colours and forms of plants. I suffer for them when drying wind tugs and torments them, bathe with them in the stillness and dampness when we have been blessed with a brief passing shower. Like watching a play we stand enthralled, knowing every scene is short-lived. Then the shears must go in, the spent beauties tidied away to make room for the next act. If we have planted with forethought there should be no shortage of performers.

Along the winding gravel path, its contours blurred by the overspill of plants edging into the gravel, there are waves of colour ebbing and flowing between plants waiting their turn to come throughout the summer months. Every day from dawn until dusk new characters unfold. The wild gladiolus, *Gladiolus communis* subsp. *byzantinus*, pleases me when still in bud. (That is the case with so many things – the promise of seeing again apple blossom, buttercups, a bush of lilac, is so much more heart-stirring when the first buds are opening, the intensity of emotion draining away at full bloom, as the colours fade.) To return to the gladiolus: spires of upward-pointing, long, narrow buds above lance-shaped leaves break the monotony of many rounded plants. As I passed them by, not one flower was showing. Within an hour or so the first satin-finished petals had unfolded, a savagely beautiful blend of purple and bronze making brilliant accents above grey-leafed acaenas and pink-flowering mats of thyme. Close inspection of the individual flowers makes me think of a paint box, or perhaps a palette, where purple and coral-pink have washed together, with one brushful of white stroked along the centre of three lower petals to guide insects to the secret heart. Seeded around the gladiolus, forming an analogous harmony, is *Linaria triornithophora*, just opening small purple-lipped flowers, from tiers of buds looking very like green-striped budgerigars, held in clusters of three or four at intervals along branching upright stems. Found in both pink and purple shades, it will surprise with jewel-like dashes of colour for weeks to come.

Elsewhere I find an extraordinary combination of scarlet *Tulipa sprengeri*, flirting around a clump of *Gladiolus communis* subsp. *byzantinus*, a most starling combination. (Christopher Lloyd would love it!) Below them a patchwork quilt of dark slate-purple *Sedum* 'Bertram Anderson' is combined with the deep rose of *Gypsophila repens* 'Rosa Schönheit'. Sprawling stems and seed heads of *Euphorbia myrsinites* try to calm the clamour while soft purple feathery seed heads of *Pulsatilla vulgaris* shiver in the breeze beside the succulent sedum. Behind them all, bold and comfortable like well-placed furniture, is a *Yucca gloriosa* 'Variegata' and an inviting mound of grey-felted *Ballota pseudodictamnus*.

Tulipa sprengeri is the very last tulip to flower, in late May. Undiscovered when in slim green bud, they suddenly appear like scarlet poppies of the field, and almost as quickly are gone, leaving slim ribbed seed pods to scatter their seed, to emerge in two or three years, a little glade of scarlet shapes. *Helianthemum* 'Tomato Red' makes a superb foil for the lily-shaped tulips. This helianthemum hangs on to its flowers rather longer into the afternoon than most, a warm, soft suffused tomato-red, its translucent petals pinned together with a minute green knob.

THREADS OF GOLD

While pinks, mauves and blues are coming fast, including various forms of catmint, thistle-like centaureas, wide carpets of thyme and many different salvias, yellow still runs like a thread of pure gold through this landscape of myriad shades and textures. Two shrubs – Spanish broom, *Spartium junceum*, and Mount Etna broom, *Genista aetnensis* – form major features, their scent filling the air this warm, dry, sunny morning in May. The Spanish broom is just opening pointed spires of bloom on stiff, knitting-needle-like stems. The odd lemon-yellow tree lupin, *Lupinus arboreus*, seeded in the background can be seen through the veil-like flower stems of the huge *Crambe cordifolia* forming a maze of stiff, branching

stems 2m/7ft plus tall, peppered with tiny green buds about to burst into a cloud of small creamy white, honey-scented flowers.

Tall candelabras of woolly-leafed verbascum crowded still with clear yellow flowers stand etched against the dark clipped leylandii hedge, or pose, isolated, on a border edge, thrilling verticals high above low cover plants, where you can admire them from top to toe and enjoy them all round, seen in a void.

Yellow is repeated in the button heads of santolina, *Achillea* 'Moonshine', the snapdragon spires of *Linaria dalmatica*, yellow horn poppy, yellow-leafed *Origanum* (*O. vulgare* 'Thumble's Variety') and, just coming now, an enchanting local native (almost lost in the wild) *Bupleurum falcatum*. Since it is a biennial, seedlings showing now along the path edge will flower next year, making delicate branching stems crowded with lacy heads of tiny yellow cow-parsley-like flowers. The whole plant, with its screen of almost leafless green stems and flower heads scattered throughout, creates a similar effect of freshness and daintiness in high summer, as does the froth of pea-green *Euphorbia stricta*, seeded among *Geranium subcaulescens*, *Origanum* 'Rosenkuppel' (yet to flower) and *Stachys byzantina* 'Primrose Heron', whose spring dress of lemon-yellow has faded now, the mature rosettes of leaves being pale grey. New yellow leaves will appear with autumn growth.

A particularly lovely shade of yellow is found in the cool saucers of the wild hollyhock, *Alcea rugosa*. Unfortunately, like all hollyhocks, it is subject to rust, which spoils the leaves, but as it is rising from behind bulkier plants the rust is not too apparent. It self-seeds wherever it can find a space, sometimes forming a group in the back of a border where it is supported usefully by neighbours, sometimes appearing among low plants on the border edge. I debated for weeks whether I should pull it out from there, but finally did not, and now I sit beside it on a low stool, writing this and enjoying it, standing proud and alone. Strangely I notice this plant has scarcely any rust. Does that tell me something? Is it more aerated at the border edge so the disease has not yet taken hold?

Phlomis fruticosa is about to release its claw-shaped yellow flowers. It makes a relaxed tumbled shape, with its tight bunched heads appearing from ground level up to 90cm/3ft or 1.2m/4ft. Prune it after flowering. You can rejuvenate a crowded or leggy bush by thinning branches and pruning worn-out pieces from the base. You will find you are removing half the foliage, but you are making space for good strong growth.

Elsewhere, the white-flowered form of saponaria has made low cushions 90cm/3ft across and about 25cm/10in high, smothered so densely with tiny white flowers they compete for attention with the bright yellow domes of *Genista lydia*. The very free-flowering sage, *Salvia lavandulifolia*, is a good foil for yellow leaves and flowers, making a ribbon of blue between the *Genista lydia*, *Achillea tomentosa* and *Allium moly*, whose green-striped fresh yellow flower heads are a special delight to me.

Various forms of wallflower, *Erysimum*, provide tones and textures of yellow to contrast with the background furniture of woolly grey ballota, unopened spikes of lavender, sword-like spikes of yucca or the yucca-like *Kniphofia caulescens*. *Erysimum pulchellum* forms a large low circle of soft growth topped with spires of acid-yellow flowers, while *E.* 'Bowles's Yellow' is more shrub-like with denser heads of warm yellow, scented flowers opening from dark purple-stained buds. This wallflower has a very long season, its first flowers opening in late March, and still looks handsome in early summer, its closely held flower stems with buds masking the lower stem, where flowers have faded and fallen. It does not appear to set seed.

Another pretty low-growing wallflower was given to me as *E.* 'Rosemoor'. Its leaves are lost beneath the crowded heads of flower, busily worked by several kinds of bee and tiny hoverfly-like creatures. From a distance the effect is of creamy yellow, but close to you can see clusters of maroon-stained buds opening into pale yellow flowers which fade to cream suffused with light purple. Nearby a fine display of mauvish pink daisies with blue and yellow eyes – *Osteospermum jucundum* – picks up some of this weird colour scheme and will go on long after the wallflowers have faded.

PINKS AND MAUVES

Perhaps the most magical effect of all in the Gravel Garden, on a sunny Sunday afternoon in early June, is created by two forms of oriental poppy. Both originated in Germany. *Papaver orientale* 'Juliane' holds salmon-pink crinkled petals above *Anthemis punctata* subsp. *cupaniana*, a tumbled mass of chalk-white, yellow-eyed daisies. Tall-stemmed black-purple onions, *Allium atropurpureum*, stand among them, repeating the knobby seed cases of the poppies. Beyond the rose-pink *Cistus*

TOP *Verbena bonariensis*

BOTTOM *Papaver rhoeas*
Mother of Pearl Group

OPPOSITE Milder winters have
encouraged us to plant more
unusual lavenders, provided the
site is sunny and the soil well
drained. *Lavandula pedunculata*
subsp. *pedunculata*, with
longer tip 'petals' than the type,
probably evolved to attract
insects to the spikes of fertile
flowers beneath. *Tulipa sprengeri*,
the last of the wild tulips to
flower, flares rebelliously in the
foreground, its slim green buds
invisible until the last moment,
when suddenly, one bright
morning, they burst open.

'Peggy Sammons', *C. × cyprius* is just opening. Tall, white, felt-coated candelabras of verbascum rise against the rich green close-clipped hedge of leylandii.

The other pink poppy, *Papaver orientale* 'Karine', is a little shorter, 75cm/30in at best. Many of the stems are shorter, self-supporting, with wide shallow cups – translucent, palest salmon-pink, lovely – nestling beside a large clump of *Stemmacantha centaureoides*, a grandly architectural plant, clothed in jagged grey leaves, white beneath, holding stiffly erect large golf-ball-sized buds consisting of overlapping silvery transparent scales. They open to release a powder-puff-like confection of finely shredded purple petals, stamens and pistils.

Centaurea pulcherrima is good for the border edge. Above finely cut grey-green leaves stand stems 45–60cm/18–24in carrying wide cornflower-shaped flowers. From silvery buff buds clothed in feather-like bracts they finally open very beautiful flowers 8–10cm/3–4in across; a rim of soft mauve tubular flowers with deeply cut petals surround the white central boss containing stigmas and stamens. No doubt these outside frilly flowers are to attract insects to pollinate the centres.

A strain of the scarlet cornfield poppy, *Papaver rhoeas*, originally developed over many years by Cedric Morris, has finally been named Mother of Pearl Group. The tissue-paper petals may be found in many shades, some single, some double, ranging from pure white thinly rimmed with scarlet to grey-purple streaked with cotton-thin red veins. They have the rapidly emerging dark stems of *Verbena bonariensis* for background. Beside them low mounds of *Lavandula stoechas* with dark violet velvety flower heads are enlivened by ascending stems of *Nepeta tuberosa* full of promise to come, the elongated flower buds (up to 60cm/24in) repeating the colour of the seed heads of *Allium karataviense*. The boat-shaped leaves are fading and drying now, but the unripe seed capsules are fleshy and handsome, a soft dull mauve colour, about 12cm/5in from ground level, each head 7–10cm/3–4in across. *Allium cristophii* is superb, standing up to 60cm/24in tall, globular heads of metallic star-petalled flowers coming through grey-blue foliage of *Euphorbia pithyusa* not yet out, but a feature nonetheless.

Further along, separated by the spreading chrysanthemum-like leaves of *Artemisia stelleriana*, is *Osteospermum jucundum*, its large clear mauve-pink daisies with bronze backs revelling in the hot sunshine. A small, pretty, new *Diascia* 'Lilac Belle' looks well in front of the Mother of Pearl Group poppies; it has shaded pink petals with dark eyes. We shall have to wait to see how hardy it is. If it behaves like other diascias, it should spread, cover its allotted space of gravel and flower well into late summer.

An addition to the campion family is *Lychnis × walkeri* 'Abbotswood Rose', barely 60cm/24in tall. It has somewhat lighter flowers than the type, although they are still a good vibrant colour. Grey ascending stems of *Artemisia ludoviciana*

PREVIOUS SPREAD Agapanthus perform well in the well-drained, arid soil and provide useful colour and form during the often dry days of August. Here Agapanthus 'Isis' is surrounded by the silvery-white velvety bobbles of Stachys byzantina 'Cotton Boll' and a sea lavender Limonium platyphyllum 'Violetta'. The lavenders and blues are picked up by a backdrop of Verbena bonariensis.

make good background with the sharp acid-green of *Euphorbia stricta* alongside. Skirting it, making good ground cover, is *Calamintha grandiflora*, just opening small, lipped tubular pink flowers.

Small cranesbills (*Geranium*) make good edging plants. *Geranium cinereum* 'Ballerina' forms small neat clumps of round frilly green leaves. Thin stems carry a long display of good-sized flowers 3cm/1½in across, heavily veined and stained, giving overall a purplish pink lace-like effect. *Geranium subcaulescens* is similar in habit with flowers of jewel-like intensity, their base colour being vivid campion-pink, with only a few dark veins leading to small black eyes.

Young plants of *Thymus doerfleri* trailing stems of woolly leaves into the pathway will next early summer carry round heads of tiny pink flowers. Meanwhile *Phuopsis stylosa* has been flaunting for several weeks crowded carpets of lax stems topped with round pin cushions studded with tiny starry pink flowers, each with long protruding anthers. Such an easy colourful plant – but it must be curbed, cut well back after flowering, to rescue whatever may be smothered beneath its ebullient growth.

Helianthemum 'Annabel' makes clusters of pink petals rather like artificial flowers made of crêpe paper, each dark petal base making shadowy depth in each individual pom-pom of flower. They tumble in front of cool grey emerging flower heads of *Stachys byzantina* 'Cotton Boll', whose velvet-leaf carpet shields rising stems of regal lilies.

Lavandula pedunculata subsp. *pedunculata* 'James Compton' is a very handsome variation of the type. Longer in all its parts, taller (60cm/24in plus), its dense dark heads are topped with long narrow wavy petals, of light purple, distinctly pinkish, perhaps bright mauve, an unpopular adjective, which is a pity, because it is a lovely glowing colour with the low afternoon sun coming through. Grandly vertical behind is a group of *Phlomis tuberosa* 'Amazone'. Strong self-supporting stems 120–150cm/4–5ft are set with pairs of dark green pointed leaves, diminishing as they reach the top. The stems are dark purple and each node carries a whorl of pale lilac lipped tubular flowers, opening over several weeks in early summer. It is a most handsome and valuable vertical, rising above low mounds of ballota, purple-leafed sage or the various santolinas. The perennial purple-leafed fennel stands further along the bed, waiting to be tangled by the viticella *Clematis* 'Étoile Violette' as it travels from its first host, the pretty lilac-pink form of *Rosa spinosissima* 'Falkland' on its way finally to the woolly spikes of *Perovskia* 'Blue Spire'. Going back to James Compton's *Lavandula pedunculata* subsp. *pedunculata*, this also is quietened and set off by the yellow-toned leaves of *Salvia officinalis* 'Kew Gold' and on the other side by *Caryopteris* × *clandonensis* 'Worcester Gold'. The purple globular heads of *Allium hollandicum* 'Purple

Sensation' look well coming up out of the matt velvety purple leaves of *Salvia officinalis* 'Purpurascens'.

Cistus 'Peggy Sammons' has such a light elegant habit, slightly felted, scented leaves and a profusion of soft rose-coloured, rose-shaped flowers. I fear it cannot be reliably hardy! But it grows so fast – 90–120cm/3–4ft high and as much across after one year – that you can afford to take cuttings and replant. Wonderful with ballota, now at its best – whorls and whorls of mouse-soft, grey-green cupped leaves, containing tiny clock faces of minute florets. Strong colour harmony is provided with spires of *Gladiolus communis* subsp. *byzantinus* and an unusual sage, *Salvia nemorosa* subsp. *tesquicola*, whose basal leaves are lost beneath close-packed stems, up to 90cm/36in tall, carrying narrow spires of beetroot-pink bracts releasing small gentian-blue lipped flowers. The colourful bracts make a display long after the flowers have set seed. At their feet *Sedum* 'Ruby Glow' spreads purple succulent trailing shoots across the gravel, interplanted with *Allium unifolium*. Like sequins, the tiny vivid pink flowers of *Dianthus deltoides* are planted to make mats, and contrasts of scale and colour among those pink- and purple-toned plants.

SELF-SEEDERS

One of the island beds has developed a mixture of blues, mauves, yellow and white in this season. I say developed, because some of the plants have placed themselves by seeding, even invading from far away, as has the lemon-peel clematis, *C. tibetana* subsp. *vernayi*. It has put itself among a huge swathe of *Bergenia* 'Rosi Klose' where its fishing-net-like stems veil the stabilizing effect of these large, simple leaves, used here and there to form a barrier or resting place for the mind and eye, among many small, fussier leaves. If you use a clematis like this, on the ground, to trail over other plants, you need to watch its progress and train the shoots where you want them to go, rather than wait till they are an inextricable bundle, like knitting wool tangled by the cat.

The lemon-yellow of the clematis's little 'lampshades' is picked up by a lovely form of Californian poppy, *Eschscholzia californica*, sent me by my first American student, Jack Henning, more than ten years ago. It self-seeds, so every year we have a little 'meadow' of this plant of heavenly colour, which waits for the sun to encourage it to open wide. Seeded among it is a curious catmint, *Nepeta tuberosa*. Standing up to 60–75cm/24–30in, stiff, upright branching stems clothed in grey, felted leaves carry large densely packed spires of blue flowers in dark purple calyces. They make much needed strong verticals among a variety of annuals in this situation. Overall, annuals don't have much chance to find a foothold where perennials, including some small shrubby plants, make good ground cover, but

OPPOSITE TOP *Nepeta tuberosa*

OPPOSITE BOTTOM LEFT
Galactites tomentosa

OPPOSITE BOTTOM RIGHT
Verbena bonariensis

in this particular area the soil is the poorest gravel and sand, facing due south, so long weeks of drought have weeded out all but the toughest. I am pleased this strange catmint has survived and colonized the area, but know if we were to have severe winter cold we could well lose it.

A strange annual, *Galactites tomentosa*, makes a feature here for weeks. It is a much admired thistle, pressed flat against the soil in spring, when the young heavily silvered leaves, intricately cut, remind me of snowflakes. It develops to make a much-branched plant still silvered, and crowded with pale mauve, fluffy, thistle-like flowers. It is removed now, having set seed, and another vertical effect is provided by the tall, dry, square-shaped stems of *Verbena bonariensis*, topped with branching heads of bright mauve flowers. They rise from mounds of santolina, *Euphorbia pithyusa* and squeaky fresh *Sedum* (Herbstfreude Group) 'Herbstfreude', whose flat heads of jade-green buds will please me for weeks to come before they bloom in September.

The centre of this bed, beneath a young tree of *Genista aetnensis*, is filled with *Tanacetum niveum*, looking like a superior form of feverfew. Each year it makes stiff, almost woody stems clothed in fine-cut leaves, bearing for weeks throughout June to early July crowds of little white daisy flowers. Backed also by *Yucca recurvifolia* (just opening now) and self-sown onopordum, it is fronted by a group of *Euphorbia seguieriana*, which I have allowed to seed. This simple but bold effect leads into a more detailed area, where *Triteleia laxa* (syn. *Brodiaea laxa*), a lovely bulb from the US prairies with wide open heads of upturned blue bells, forms a deep pool of colour, veiled with self-sown bupleurum, while tall spires of *Gaura lindheimeri*, pink-washed buds, moth-like white flowers, hover above them and thymes roll out a carpet beneath them into the gravel paths. Madonna lilies, *Lilium candidum*, have taken several years to decide if they would live or die, but now this summer they have delighted me. Not just one struggling stem, but several from the original bulb, and good-quality flowers, although the foliage has gone. I'm told if you grow them from seed they are less likely to be virus infected. After 125mm/5in of rain this summer, maybe we will have the chance to collect seed from well-nourished plants.

The early euphorbias are starting to dim now, as they set seed, but *Euphorbia corallioides* is just beginning and will maintain that wonderful spring-green, its far-flung seedlings having tucked themselves into the smallest space, spreading wide heads of fresh yellowy green, often providing a cooling effect just where it is needed. The unwanted seedlings are easy to recognize and pull up if they are smothering something more important.

In the autumn we scattered the seed of the annual *Omphalodes linifolia* in seemingly bare spaces. Now in flower they froth like a wave's edge breaking along

our gravel pathways. Each plant makes a neat little bushlet, about 30cm/12in tall, covered from top to bottom with narrow, glaucous-grey leaves, crowded with starry white flowers. In one group they repeat the white daisies of *Anthemis punctata* subsp. *cupaniana* and *Helianthemum* 'Wisley White', and the same colour is repeated in the background with huge bouquets 2m/7ft tall of *Crambe cordifolia*. All this white is relieved with the light purple of *Lavandula pedunculata* subsp. *pedunculata* 'James Compton' and repeated in more *Allium hollandicum*, while *Alstroemeria ligtu* hybrids and clumps of opium poppy stand by to take their turn.

As the season progresses some of these self-seeders have to be removed. Some should have been removed earlier, but sometimes it is easier later either to pull them out when there is enough to handle, or to make the decision to let them be. It becomes obvious when there is too much of something and especially so this season after 100mm/4in of rain, followed by unaccustomed good growth.

Helianthemums tend to look their best in the morning, especially in hot bright weather. Each flower lasts one day, like cistuses. But there is a copious supply of buds, so the colour effect lasts for several weeks. There are many named varieties, both single and double. I tend to prefer the singles, but there is both a pink and a lemon-yellow double which are very likeable. *Helianthemum* 'Rhodanthe Carneum' (syn. 'Wisley Pink') is deservedly a favourite, smothered in rose-pink flowers with flushed yellow centres. If pruned back well after flowering the plants quickly regrow into tidy clumps, several closing together to cover the soil. White love-in-a-mist, *Nigella damascena* 'Miss Jekyll Alba', is seeded through them together with *Allium schoenoprasum* 'Forescate', a pretty form of edible chives, rose-pink in colour instead of mauve. All this is in an area which was blue in late February with scillas and ipheions. A self-sown aquilegia in palest shell-pink stands defiantly above this group as if to say, 'You dare!', knowing it is in my mind to pull it out since aquilegias associate, I think, in cooler situations; but my heart stays my hand until the flowers are over.

ROOM FOR GIGANTIC PLANTS

In the wide border at the far end of the garden, towards the wide wooden farm gate, early June brings a combination of yellows, blue and white. *Achillea* 'Moonshine' and 'Taygetea' both provide fresh gleaming colour, with cloudy blue *Nepeta* 'Six Hills Giant' for contrast. But since the border is 10m/30ft across, here there is also room for gigantic plants.

Besides *Cistus* × *cyprius*, *Abutilon vitifolium*, *Lavatera* × *clementii* 'Bredon Springs', various forms of eucalyptus tower above the clipped hedge, breaking the horizontal line. Together with giant herbs and grasses, onopordums make wonderful carved outlines; verbascums join them with softer wool-coated clusters of buds along their candelabra-like flower stems. Variously known as golden oats and giant feather grass, *Stipa gigantea* from Spain is more than 1.8m/6ft high and creates a shower of needle-like stiff stems, every polished bronze awn catching the sunbeams. A selected form of cardoon, *Cynara cardunculus*, chosen for its especially fine-cut silvery grey leaves, already stands 1.8m/6ft tall, making a majestic contrast to many fine-leafed plants nearby. Close to it but not in the least intimidated, another impressive plant is *Crambe cordifolia*, its huge panicles of white, honey-scented flowers almost concealing the yellow tree lupin, past its best now, which lit up that part of the garden in May. Quite near the front of the border is another crambe, *C. koktebelica*. I value highly this plant. It is not nearly so demanding of space, being less than half the size, l–2m/3½–7ft, and enchanting in effect. It has deeply cut leaves and large gypsophila-like heads of white flowers (coming in as *C. cordifolia* is beginning to fade). When it is finished and cut down, late bulbs like galtonias come up to take its place.

Yellow forms of *Eremurus × isabellinus* Shelford hybrids stand tall and slender like altar candles above this wide range of textures and colours. But *E. himalaicus* is the first of the foxtail lilies to flower, with immense cylindrical spikes of pure white flowers, dramatic verticals, outlined against the clipped leylandii hedge. *Eremurus robustus* bears rose-pink flowers a little later. Both can be 1.5–2.5m/5–8ft tall. These lilies come from hot and dry regions of the Himalayas and Turkestan and are unreliable in areas where late spring frosts can destroy their lush foliage and thus starve the strong roots needed to support such a spectacular flower spike. Planted in the back or centre of borders, the gaps left, when the eremurus leaves have disintegrated, are hidden by later developing plants growing around them.

It would be unfair to leave out a quiet but distinctive bulb, *Ornithogalum pyrenaicum*. When it is planted in groups, its bare 90cm/3ft stems, topped with slender spires of six-petalled starry pale green flowers, add a peculiar delicacy among the anonymous plants beneath them quietly waiting their turn to rise up and show their worth in the weeks to come.

BACKGROUND TREES AND SHRUBS

Despite being stressed at times by drought over the past seven summers, the carefully selected trees and shrubs (listed on pages 182–3) forming the background and permanent furniture of the Gravel Garden appear suddenly to have put on surprising girth and height, creating an undulating landscape, especially where the borders are wide – as much as 10m/30ft in some places.

Abutilon vitifolium 'Album', at 5m/16ft almost as tall as the clipped leylandii hedge sheltering it from winter winds (and 3m/10ft wide), is alive with large chalk-white flowers quivering like butterflies against a backcloth of richly green, pointed, vine-shaped leaves. Every terminal shoot on this huge bush bears large clusters of buds, some round and tight, others lengthening like furled umbrellas waiting to flare open wide five slightly wrinkled petals held together with a small boss of cream stamens, looking like huge single roses. Rising up alongside is a slim pencil juniper, *Juniperus scopulorum* 'Skyrocket', not yet quite tall enough to make the apex of a triangle formed by the abutilon, the holly-green-leafed *Cistus ladanifer* and the Spanish broom, all suitably bulky in scale, while the wonderfully jagged silvery grey leaves and the noble flower stems of the cardoon, *Cynara cardunculus*, provide both a change of texture and tone. Bringing the picture down to mid-border and repeating the light grey tints are *Brachyglottis* (Dunedin Group) 'Sunshine' and grey and green forms of cotton lavender, *Santolina chamaecyparissus* and *S. rosmarinifolia,* now completely restored with fresh young growth. After looking so tatty in late winter I thought we should replace them, but didn't get round to doing so!

I stand, on a soft summer's morning, holding my breath with wonder, beneath the huge white-flowered abutilon, looking so fresh and perfect, knowing how ephemeral it can be in my garden, at risk from severe frost and drought. Elsewhere in the Gravel Garden I have lost one through drought. This one has come through three taxing summers, but is now, I fear, much too large to survive without irrigation if we have, as is usual, many weeks of drought and heat to come. When its flowers have faded, or before if necessary, it will be pruned to ease the burden, reducing its bulk by at least a third.

Much smaller (so far) is *Abutilon × suntense*, placed about 4m/13ft away from the roots of the leylandii hedge. Although smaller in all its parts (now about 1.5m/5ft across and through), its effect is as dramatic, being covered in trusses of nodding flowers, violet-blue with a faint tinge of pink washing through hair-fine veins, the overall violet tone deeper at the edges of the wavy petals, paling towards the centre. Nearby large rose-pink petals of *Cistus × purpureus* hold their own, while a pink-flowered lavatera stands by, in tight bud, ready to provide long, graceful boughs of blossom for weeks to come. Two fine verticals enhance this group, already lifting the eye above surrounding mounds of santolina and ballota. They are giant feather grass, *Stipa gigantea,* already forming a golden haze, while wickedly spiny, but gorgeous as contrast, is *Onopordum acanthium*, whose broad, deeply cut leaves tipped with spines look as if cut out of stiff felt.

PROBLEMS WITH CONIFERS

We planted the leylandii hedge, which makes such a feature in the Gravel Garden, about thirty-five years ago. It is approximately 300m/1,000ft long and extends the whole length of the car park and nursery stock areas. As well as making an important background, the hedge provides some shelter from winds out of Russia, which sweep across East Anglia, especially cold in the winter. (I do sometimes point out, however, that the Gravel Garden can only be partly sheltered by the hedge, and that if you stand in my 'dried-up river bed' – the central winding gravel walk running due north and south – you are still in a wind tunnel.)

Soon after we had embarked on the Gravel Garden, after the particularly hot and dry summer of 1995, we noticed rusty brown patches appearing here and there in the hedge. The next year these patches enlarged at an alarming rate. From rusty brown they turned to dull, dusty brown. A brush of the hand, and the completely dead foliage fell to the ground revealing bare stems.

The original trees had been planted 1.25m/4ft apart. In some instances the whole tree died from top to bottom. But we noticed strong young shoots spreading from neighbouring unaffected trees across the dead gaps. The hedge was due for its annual clip. Would it be possible, I suggested, instead of clipping those new shoots, each about 30cm/12in long, to tie them in; train them to grow across the dead areas? The task may have seemed like emptying a pool with a teaspoon, but Keith Page, my technical manager, began immediately, using short lengths of soft hessian string and ensuring the shoots faced upwards if possible, horizontally where necessary but never downwards.

There is always a tendency with hedge cutting, as with lawn mowing, to make a thorough job, to cut too close. In both instances this is harmful, especially in hot, dry conditions. From then on, we cut less severely, usually in the first week of July. Over the years many of the gaps have completely greened over by tying in young shoots to become strong horizontal branches. But we were still worried because few conifers (yew and swamp cypress are exceptions) make new growth from bare wood. Would we lose the entire hedge? What was the cause? I felt several consecutive dry summers with abnormally high temperatures (25–30°C/77–86°F) must have had some effect, but talking with visitors I learnt that there are both a pest and a disease attacking conifers around the temperate world. So I wrote to Dr Chris Prior, Head of Plant and Disease Science at the Royal Horticultural Society's Gardens at Wisley, describing our problem. I quote his reply: 'The most likely causes are either infection by the fungus disease known as *Coryneum* canker, or infestation by conifer aphids, in the genus *Cinara*.' However, both Dr Prior and David Pycraft, Horticultural Adviser to the RHS, believed the deterioration followed stress due to drought conditions, particularly since our 'soil' is light, sandy gravel.

Not altogether convinced, we started spraying the hedge with deterrents – the first application during the second week of June, and then immediately after cutting, using malathion for juniper scale and iprodione (as Rovral Wettable

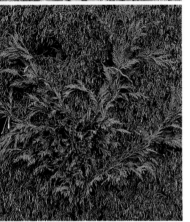

TOP *Hedera helix* 'Glacier'

BOTTOM Young live shoots have been trained across the dead face of the hedge.

OPPOSITE A principle of planting used here is to use wandering plants to flow in and out between more stable neighbours, such as ballota or santolina. *Euphorbia corallioides* and love-in-a-mist (*Nigella damascena*) need watching lest they become too invasive. The superb thyme in the foreground is *Thymus longicaulis*.

Powder) as a fungicide. I confess I still did not feel totally confident of our timing, especially with the fungicide. As you may already know, we need to protect fresh growth with this material, so that attacking spores are killed on contact. To counteract stress brought on by poor growing conditions, that is, gravel soil, we applied foliar feed, Sinclair-Sangral Soluble fertilizer (3:1:3).

The results of all this appeared to be rewarding. Where the hedge had grown back to look as good as new by tying in neighbouring shoots over several years, there was no more damage – until the summer of 1998, when new patches of rusty brown suddenly appeared. Gardeners and photographers asked anxiously what we could do about it. At the same time, I became concerned to see some dieback of foliage at the base of several young specimens of *Juniperus scopulorum* 'Skyrocket', planted in the island beds. This problem, I saw from the RHS Entomology Advisory Leaflet No. 46, is caused by the juniper aphid (*Cinara fresai*), a species of North American origin but now widespread in gardens in southern England. It forms dense colonies on young shoots and can be found from May to October (not good news!). Then in the spring of 1999 a visitor who had an equally long length of leylandii hedge and much more experience of this problem confirmed that the damage to our hedge was caused by aphids. He thrust his hands deep into the rusty brown foliage and drew them out to show us they were wet with aphids' secretions! Ugh! We were glad to follow his recommendations.

In the summer of 1999, we sprayed the hedge twice, at the beginning of June, and again a month later, in early July. We again used malathion, but more systematically, using a pressure lance, fitted with a cone. (A knapsack sprayer, we were told, is not adequate.) To kill all the aphids breeding and feeding inside its warmth and shelter, the lance, powered by the tractor, must be pushed well into the centre of the hedge with a push-and-pull action, at appropriate distances apart to ensure every area is covered. Malathion is a form of organophosphate, which means the operators must be fully dressed in protective clothing. Applying it is a slow and arduous job, but must be done thoroughly: it is a waste of time if any infected part is omitted.

Although dismayed by the new infection, we now know what to look for, and shall continue to look carefully for signs of aphids, by opening the hedge and pushing a hand well into any freshly browned foliage where the aphids live in the warm, eating their way from inside. We hope we can contain them by applying insecticide efficiently and by cutting less severely.

ELEGANT ALLIUMS

For weeks, from late spring until late summer, many different alliums thrive in the Gravel Garden. Planted initially as small groups of individual bulbs, they have increased by division or seed to create natural-looking drifts, the tall ones poised high on bare stems, the shorter ones pushing their round heads, like golf or tennis balls, through low cover plants. *Allium karataviense* and *A. oreophilum* look like this. *Allium hollandicum* forms a dense globe-shaped head, in various shades of purple, some almost lavender-coloured, the deepest an almost blackish purple. All thrust tall stems through mounds of purple-leafed sage, their shrivelled leaves buried in late May in a foreground of *Sedum* 'Matrona', gauzy clouds of catmint in flower and velvety carpets of *Stachys byzantina* whose leaves, stems and flower buds are completely whitened with wool. A stately *Verbascum bombyciferum* stands at the edge of the path, its bold shape and pale colour acting like a full stop, a point to rest the eye among a wealth of detail.

Colonies of *Nectaroscordum siculum* subsp. *bulgaricum* (we used to call it *Allium bulgaricum*) are at their best towards the end of May. On strong, slender stems the elegant heads of bell-shaped flowers are downcast until fertilized by bees, when they rise erect to form clusters of seed cases looking like fairytale castle turrets. They increase readily, so it is not difficult to have dozens of stems swaying in a soft breeze. The colour is indescribable. In *N. siculum* the basis is alabaster cream, heavily overlaid with plum, producing from a distance a curiously beautiful brownish pink shade. *Nectaroscordum siculum* subsp. *bulgaricum* has much less plum, its petals striped alternately in plum, cream or green creating a paler effect. Whether as a single rogue emerging where a seed has shot, or drifting in colonies through tolerant neighbours, these visions of stateliness and subtle beauty are unforgettable. But do remember to remove the seed heads (after giving enough time for their effect) before the seeds have dropped.

Allium unifolium stands about 25cm/10in tall, with upturned umbels of star-shaped flowers, bright cyclamen-pink. Increasing well, it now makes a bright patch of colour through a carpet of pewter-leafed *Acaena saccaticupula* 'Blue Haze', a modest plant, crowded with round burr-like flower heads distinguished only by there being so many of them. Curiously brown on thin brown stalks, they create an unexpected dark effect beneath the bright allium, a shadow weaving among the pale stones of the gravel pathway.

I am thrilled to see seedlings of *Allium cristophii* flowering now from parents planted originally six years ago. They take several years to make a flowering-sized bulb, unlike culinary alliums. In one of my island beds I have counted over thirty heads, pushing through and above the lovely blue flax, *Linum narbonense*,

PREVIOUS SPREAD One of
the most exciting times in the
Gravel Garden is when the
alliums go marching through.
Allium hollandicum (syn. *A.
aflatunense*) comes in various
shades of purple, the deepest
called *A. h.* 'Purple Sensation'.
The fiercely yellow flowers of
Genista lydia and the cool
yellow leaves of *Caryopteris* ×
clandonensis 'Worcester Gold'
make lively contrast, while
the velvety purple shadows of
Salvia officinalis 'Purpurascens'
complement the sombre tones
of the darkest alliums.

OPPOSITE This picture
shows the two forms of
nectaroscordum mingled
together. *Nectaroscordum
siculum* has translucent bells
stained wine-red and green.
Nectaroscordum siculum subsp.
bulgaricum has cream flowers
softly shaded with green and
pink. I like the effect of all these
smooth bare stems, including
those of *Allium nigrum* in the
foreground emerging from a
carpet of *Stachys byzantina*
'Cotton Boll'.

while the floor beneath them is scattered with dark seedlings of *Euphorbia dulcis* 'Chameleon' and the purple succulent-leafed *Sedum telephium* subsp. *maximum* 'Atropurpureum'. *Verbena bonariensis* (tall) and *Verbena rigida*, both with bright purple flowers to come, are rushing up among them and will enhance the group when the alliums remain good as bleached seed heads. It was here in spring that David Ward and I planted *Iris* 'Black Swan' among this group. Although only producing a single stem each this summer, the iris's dark velvety tone and contrasting shape already add an exciting element.

I can't make up my mind about *Allium nigrum*. Why *nigrum*? So named, I am told, because some populations have black ovaries. Mine have green ovaries and produce flattish heads of white flowers, each individual floret pinned with a green eye. It makes lusty clumps of bulbs, so quantities of flower stems are striking in May, against a dark background with flowers at various heights. Today I am admiring the effect. But this bulb is almost too successful – needs to be repressed from time to time, certainly dug out and removed where its lush leaves in spring threaten to choke a neighbour. Yet the flower heads can and do make an interesting effect, as when their strong green stems grow in a colony in front of the handsome, leafy vertical *Phlomis tuberosa* 'Amazone'.

Most allium leaves can be a problem. They need to be planted among or behind robust growers like *Euphorbia characias* subsp. *wulfenii*, forms of *Salvia officinalis*, santolina, ballota, lavender or even through the open-growing *Cistus* 'Peggy Sammons', so you are unaware of their fading foliage. But take care not to pull up the foliage of *Allium sphaerocephalon*, mistaking it for grass. By late summer these inconspicuous leaves will have died down, when the plum-shaped, wine-purple flower heads emerge on thin stalks 75cm/30in tall – a pleasant surprise contrasting with silver foliage plants. I think they look best when allowed to seed, creating an almost meadow-like effect, weaving through low cover plants.

HIGH SUMMER'S GAUZY VEIL

PREVIOUS SPREAD A much admired hardy-annual thistle, *Galactites tomentosa* along with the small white daisies of the feverfew-like *Tanacetum niveum* have seeded around and about more permanent plants. Along with the lacy Love-in-a-mist, *Nigella damascena* 'Miss Jekyll White' these annuals will soon be removed to prevent excessive seeding. A lone, self-sown *Verbascum bombyciferum* comes into flower, repeating its twin vertical in the background.

The Gravel Garden, though still full of colour, thanks to overcast skies and cool temperatures, has its July look, very different from the lush freshness of June. Overall it has a gauzy, veiled appearance. You can view individual plants or vistas through layer upon layer of the stems of drier, less leafy plants, all much taller and tougher than they were in June.

During June 1998, we were lucky to have 130mm/5in of rain, which had a freshening and revitalizing effect (after a dry May) and was a joy, evident everywhere, in gardens and countryside. But now, towards the end of July, scarcely 25mm/1in of rain has fallen in the past four weeks. Even so, the beds are filled with flowers and foliage. There is the clear yellow of achilleas and anthemis, while the salmon-pink of diascias, which dominated in early July, is giving way to a hazier scene with purples and mauves, accented here and there by the slim creamy yellow pokers of *Kniphofia* 'Little Maid' and the lemon-yellow single hollyhock, *Alcea rugosa*.

Many shades and tones of yellow sparkle still. Yellow is so heartening when the sky is grey and threatening, yet no rain falls. It always surprises me, the violent aversion some people have to yellow in the garden. (They are usually into blues and mauves, or perhaps, more fashionably, cream and apricot.) Admittedly there are strident yellows I consider carefully before placing – or even reject if they are overdeveloped cultivars – but, surprising as it may seem, I find white – lovely as it can be in individual plants – more difficult if not used thoughtfully. It draws attention to itself so assertively, whereas yellow can be found in many shades and tones, to lighten but not frighten.

Still dominating the scene and scenting the air in the Gravel Garden are several tree brooms, the Mount Etna broom, *Genista aetnensis*. The original one I planted thirty-five years ago appears to have darker yellow flowers than those planted only seven years ago. Yet there seems no difference when the little pea flowers are picked and laid side by side. I think the difference in tone is an optical illusion, caused by there being, as yet, fewer flowers and more fresh green growth on the young trees. These are now about 3m/10ft tall, already forming fountains of flowers on arching boughs. *Spartium junceum*, with larger, yellow and scented pea flowers, is spikily upright in form. Pale greenish yellow is found in the shallow saucers of the wild yellow hollyhock *Alcea rugosa*, while an even more greenish yellow is found in *Euphorbia seguieriana* subsp. *niciciana* and the pretty umbellifer *Bupleurum falcatum*.

The bright brassy foliage of *Origanum* 'Norton Gold' makes a good border edge, drawing attention from cut-down clumps of *Euphorbia epithymoides*. It is not essential to cut down this plant, but in dry conditions, such as we can rely on here, it can be disfigured by mildew. After collecting seed in June, we cut the already infected stems to the ground. The plants look horrid for a while, but now the bare stumps are crowded with tiny new shoots. In a month's time or less, they will have made clean mounds of fresh foliage and usually flower a second time before the frosts take them down.

More soft yellows come from daisy flowers. I love the low-growing *Anthemis tinctoria* 'E.C. Buxton' with simply hundreds of creamy yellow daisy flowers smothering finely cut grey leaves. But I have to admit my poor thin soil tries hard the patience of this much valued plant, which flowers for months elsewhere in more retentive soil.

The santolinas have made large, tough, bushy plants, springing from a woody base, their filigree-grey woolly foliage embroidered over now with round bobbly flower heads, in both pale and deep yellow shades. After flowering in July, they should be pruned to a good shape, cutting much of the current season's growth. This mid-season cut makes densely furnished plants to go through the winter.

Walking through the Gravel Garden on a warm hazy Sunday morning, I am surprised to see *Hippocrepis emerus* (syn. *Coronilla emerus*) is almost as full now of little yellow pea-like flowers as it was in April. What a delight! So fresh here in the background, where most of its neighbours now provide only shape, tones and textures of green – plants like *Cistus ladanifer*, *Berberis* × *stenophylla* and the grey 'aluminium'-leafed *Atriplex halimus*. Crowds of butterflies, white especially, flutter through this landscape of 'hills and valleys' formed by the shapes and placing of plants.

GOOD NEIGHBOURS

But this is not just a yellow garden. Indeed not! There are stands of purple-flowered *Verbena bonariensis,* cushions of pink thrift *Armeria maritima* 'Düsseldorfer Stolz' (syn. Düsseldorf Pride) and a second flowering of rose-pink bells on *Bergenia* 'Morgenröte' (syn. Morning Red), lovely with its fresh rounded green leaves nudging against a velvet carpet of silvery grey lambs' ears, *Stachys byzantina* 'Silver Carpet'. Blue is fading from the lovely groups of the bell-flowered *Triteleia laxa* (syn. *Brodiaea laxa*) as it starts to make chaffy seed cases, but appears elsewhere with steel-blue thistle-like shapes of several different eryngiums, all suited to dry conditions. Lavender flowers are filled with the contented humming of bees where I sit on a curved wooden bench, while up and down my dried-up river bed clumps of *Agapanthus campanulatus* introduce strong tones of blue.

Now, late in July, I sit writing on my little canvas fishing stool and look into a part of the landscape where earlier cistus and abutilon thrilled me in June. There is little flower colour overall, but a satisfying fullness of foliage, of variation in

TOP *Nepeta* 'Six Hills Giant', *Allium cristophii* and *Stachys byzantina* 'Cotton Boll'

BOTTOM *Phlomis tuberosa* 'Amazone'

texture and form. At my feet *Achillea tomentosa* spreads a tatty carpet of woolly leaves, silvered with fine hairs as protection against drought. It was damaged by last summer's drought, but creeps forward now into fresh 'soil'. The 'ragged' effect does not offend me since there is no bare earth, only the stony effect of the gravel mulch. Seeded into the spaces are a thousand stems, like a little meadow of a small, delicate onion, *Allium carinatum* subsp. *pulchellum* f. *album*, just opening its first white bells. Also threading its way through the achillea is the blue-stemmed, blue-flowered *Eryngium bourgatii*, dressed in handsome, deeply cut and curled leaves, all marked with wide white veins, all very prickly, well constructed to withstand drought.

This almost cameo-like border edge leads the eye restfully into an undulating landscape achieved with bold groups of a few ground-covering shrubs. First it is embraced by a large curving sweep of grey-leafed *Salvia lavandulifolia* (a haze of blue flowers in May and June), which in turn is backed by one large, extra-'white'-leafed phlomis surrounded by several plants of blue-grey *Euphorbia characias* subsp. *wulfenii*, all resting now; then, lifting the eye and reaching for the sky, there is a self-sown group of yellow-flowered verbascum 1.5–2m/5–7ft tall. It must be a cross because it has the height of *Verbascum bombyciferum* and the dark maroon eyes of *V. chaixii*. Finally, a free-flowering form of the dune-binding grass *Leymus arenarius* with its stiff, wheat-like flower heads, tall stems and ribbon-like leaves, grey-blue in colour, rises outlined against the dark green leylandii hedge.

Elsewhere in the Gravel Garden are a few more groups looking attractive now, whether through colour or shape, or both. *Sedum* 'Matrona' backed by tall seed heads of *Phlomis tuberosa* 'Amazone' (recently stripped of its fading foliage), mingling with *Allium cristophii*, now in seed but still handsome, supported by *Bergenia* 'Abendglut' with *Thymus* 'Silver Queen' in flower now, as path edging. *Calamagrostis* × *acutiflora* 'Karl Foerster', a beautiful tall grass, is also flowering, its feathery heads stained purple-buff, repeating the dark tones of the sedum, phlomis and allium.

Seeded around and into verbena and *Gaura lindheimeri* is *Bupleurum falcatum*, creating a low haze of lace-like yellow umbellifer flowers. Providing substance are the yellow-leafed *Caryopteris* × *clandonensis* 'Worcester Gold' and *Sedum spectabile* 'Iceberg', quiet companions waiting their turn to perform in early autumn. The caryopteris are only just opening so will continue in flower well into September. I think I prefer the normal grey-leafed types (some have darker blue flowers) but 'Worcester Gold' is very attractive, with lime-green leaves, making a good foil for the lighter blue flowers – or so I thought till I compared the two. It is hard to tell the difference – but, as one finds often when handling colour, neighbouring tints can give different values to the same colour. If you select different-coloured mounts for a picture they will bring out different tones and appear to make the picture warmer or cooler depending on which colour you choose to emphasize.

In my little maritime area it is not possible to emulate the effect of plants in the wild growing on sand dunes in Norfolk or the stony hectares of Dungeness, but I have grouped together a few. At the path edge is a double form of the sea campion, *Silene uniflora* 'Robin Whitebreast'. Its heavy flowers, resembling a double pink, flop among the stones, while *S. u.* 'Rosea' is similar to the sea campion, except it is stained pink. A selected form of sea thrift, *Armeria maritima* 'Düsseldorfer Stolz' (syn. Düsseldorf Pride), makes a show of light wine-red flowers for weeks on end, while the yellow horn poppy, *Glaucium flavum*, sprawls across the path dressed in sea-green jagged leaves. It still carries curiously prickly buds, wide open yellowy orange silk-textured petals and long, rapier-like curving seed pods.

The coveted native sea holly, *Eryngium maritimum*, has taken several years to establish, to send its deep searching roots into my stony subsoil. It is not an easy garden plant, preferring the deep pure sand of wind-formed dunes. Easier to grow, and remarkable in all its parts, is *E. variifolium*, with leaves and bracts so finely divided they appear like spikes cut out of shimmering metal, each stem topped with a thimble-sized cluster of deep blue flowers. To add substance to this group I have backed it with sea kale, *Crambe maritima*, which grows magnificently on the pebbly beaches of Dungeness. In spring it bursts through the bare soil like clenched fists, tightly crumpled bundles of violet-purple leaves, which gradually expand into large glaucous-grey, wavy-edged, tongue-like shapes, followed by large heads of cream crucifer flowers on short stems. When these fade, we cut all down to encourage new, handsome leaves to appear, as now, making bold contrast.

Anthemis tinctoria 'E.C. Buxton' mingles its pale lemon-yellow daisies with the hair-fine foliage and feathery plumes of the American grass *Stipa tenuissima*, while behind them stand deep blue *Agapanthus campanulatus* 'Cobalt Blue'. We have grown this and other named forms of *A. campanulatus* for many years without frost damage. They need several years to build substantial clumps, when the flower effect is thrilling.

A screen of *Verbena bonariensis* on the curving edge of an island bed has *V. rigida*, low growing and deeper violet in tone, at its feet. This plant has been traditionally grown as a bedding plant. It is not reliably hardy in better soils, but in my well-drained gravel soil, and with milder winters, it disappears, but returns from underground shoots. Mingled among those two are the heart-stopping blue flowers of the special flax, *Linum narbonense*. It does not set seed for us (as

OPPOSITE LEFT In early summer the Mexican feather grass, *Stipa tenuissima*, produces an endless display of silvery green flower heads above soft clumps of hair-fine foliage. The overall effect of delicate plumes and foliage waving in the slightest breeze is a delight. As the flower heads mature they bleach to straw colour. If cut down, a fresh wave of young growth is encouraged while a rash of seedlings is minimized. In general I avoid free-seeding grasses, but this one is too attractive to stick to rules. Its young are easily recognizable – and not difficult to dislodge. Daisy flowers, *Anthemis tinctoria* 'E.C. Buxton', look well with grasses, while Miss Wilmott's ghost, *Eryngium giganteum*, seeds among them.

OPPOSITE RIGHT This picture illustrates the importance of a good impact plant. *Verbascum bombyciferum*, handsome from top to toe, acts as a full stop at the end of a sentence. It arrests the eye. Remove it and the scene could be a fairly confused muddle.

does *L. perenne*) so must be grown from cuttings which do not root readily, so invariably it falls into the category we label 'signed copies of limited editions'. Threaded low through all these beauties are seeded a dozen or so globe-like gauzy seed heads of *Allium cristophii*, while another group we always grew as a greenhouse specimen, *Tulbaghia violacea*, makes a foreground with fresh-looking lush clumps of narrowly strap-shaped leaves bearing a long succession of pale lilac, allium-like flowers on slender stems. A much needed touch of dark foliage and flower is provided with *Sedum telephium* subsp. *maximum* 'Atropurpureum', while the plum-coloured, plum-shaped heads of *Allium sphaerocephalon* always come as an unexpected bonus. The background to these plants, and centre of the island bed, is made with the pencil-shaped *Juniperus scopulorum* 'Skyrocket' with a great clump of Spanish bayonets, *Yucca gloriosa*, at its base. These hard yet needed outlines are softened and enhanced by viewing them through a tall fan-shaped stand of grassy flowering stems, possibly *Stipa splendens*, but we are not yet certain of the name.

HANDSOME VERTICALS

The widest border in the Gravel Garden is near the entrance gate backed by the leylandii hedge, nearly 10m/30ft from front to back. It is almost pure sand and gravel (drains like a colander), but, with an informed choice of plants and a thick mulch of straw where it does not show (from middle to back), it now presents a closely covered landscape effect, of rounds and mounds of varying height creating little 'valleys' for the eye to follow foraging bees and butterflies, while stately verticals paint the sky.

The car-park entrance has been guarded for months by the 2m/7ft tall candelabras of *Verbascum bombyciferum*. Now at summer's end a few yellow flowers still decorate the knobbly seed heads. Shining sweeps of bergenia and woolly carpets of *Stachys byzantina* lead into the first island bed, dominated now by a central group of *Verbena bonariensis*. The rigid, branching, four-sided stems need no staking. Standing over 1.25m/4ft tall they are crowded almost from ground level with flattened heads of tiny bright-mauve tubular flowers. Here they make an eye-catching feature in themselves, and also form an attractive screen through which to view a distant border, or a flat sweep of gravel, depending where you stand. Elsewhere, their upturned heads of tiny lilac-mauve flowers mingle with *Gaura lindheimeri*, its graceful branching stems alive now with fluttering white flowers reminding me of moths in the twilight.

A fine vertical is *Eryngium eburneum*. It introduces structure rather than colour, with small thimble-sized clusters of greenish white flowers that form a narrow wand on 1.25m/4ft stems waving above the basal clusters of narrow

pineapple-like leaves. Another favourite vertical is the Russian sage, *Perovskia* 'Blue Spire'. Its white-felted stems make a hazier greyer effect than *Verbena bonariensis* with its forest of dark green stems. Bumblebees and honey bees work tirelessly on the sage's tiny tubular flowers, as they do also on the caryopteris planted in front.

It is good sometimes, perhaps in low evening light, to take my stool and settle in an unexpected part of the garden, to sit and contemplate a piece of planting I normally pass or drive by. An example of this is the shrub border along the drive edge, which forms the west boundary of the Gravel Garden. Here the soil is less stony, slightly more fertile. Against a small group of dark green cypresses are outlined two bushes of *Ligustrum ovalifolium* 'Aureum', beautiful all the year round but, pruned last winter, outstanding now with fresh growth and lemon-yellow young leaves. These two woody plants make a good background for at least fifty magnificent flowering stems of *Acanthus spinosus*. They stand 1.5–1.75m/ 5–6ft tall, stout, columnar stems carrying huge spires of hooded flowers, with curved purple bracts, each sheltering one white petal, which protrudes like a tongue or frilly petticoat, all protecting the stamens and stigma hidden inside. The overall effect is somewhat dark and mysterious, interesting against the yellow privet. The deeply cut dark green leaves are somewhat disfigured this year with

powdery mildew, but this does not affect the display. From time to time deep digging is needed, in winter, to check this colonizing plant, but where space needs filling and warm summers are reliable it is both spectacular and architectural.

Perhaps the most dramatic vertical this summer has been *Ferula communis*. As I have said, it must be planted young, about the size of the pencil I write with, from a 'long-tom' pot. Remember to mark the spot because it will take two to three years to build a root system capable of supporting the purple-tinted stem, thick as my wrist at the base and standing about 3m/10ft tall, holding aloft against a blue sky upturned branches of fennel-like yellow flowers in flat heads. Emerging above all else, they come as a breathtaking surprise. Don't plant them in a wind tunnel. One of mine, just setting seed, was felled recently at the base by a particularly ferocious gale.

Tulbaghia violacea looks and smells like an allium. (Normally I do not notice the onion-like smell, but on cold, damp autumn mornings the scent can be quite strong.) It has narrow, strap-shaped evergreen leaves and, on 38–45cm/15–18in stems, carries loose heads of lavender flowers opening from darker buds. Not dependably hardy, it survives mild winters here, but to be safe we pot up a few clumps to overwinter in a frost-free place – then we can divide them like chives and plant them out in spring. They will increase and produce flowers throughout the summer and into autumn.

SPREADERS AND DOT PLANTS

I like to use occasional small dot plants as features to add liveliness to a flat expanse of carpeting plants. I planted several bulbs of *Lilium formosanum* var. *pricei* in a sea of blue *Mertensia maritima*. They have flowered well, the long white trumpets, so oddly proportioned on their short stems, standing just above tiny blue flowers. The mertensia grows wild (when it can still be found) on sheltered beaches in the Shetland Islands. Like *Eryngium maritimum*, which grows on our Suffolk sand dunes, it doesn't always take to garden cultivation. It may not take to me, but so far it appears ideally suited, spreading wide mats of prostrate stems closely set with blue-grey wax-coated leaves. Each axil carries a cluster of small bell-shaped blue flowers, opening progressively as the stems lengthen, so continuing the display for weeks, well into August. Seed is freely produced, so if my parent plants succumb, I feel certain there will be youngsters tucked down among the stones. The *Diascia vigilis* makes a pale salmon-pink wash of colour alongside the powdery-blue of the mertensia.

I planted groups of *Gypsophila* 'Rosenschleier' (syn. *G.* 'Rosy Veil') on the edge of the long winding gravel walk, in threes or fours. One plant would have sufficed, since each plant has made a cover of at least 90cm/36in across – forming soft, cumulus cloud shapes of fine wiry interlacing stems buried beneath a froth of palest pink semi-double flowers.

Nepeta tuberosa is over now but its candle-shaped seed heads formed by dense whorls of purple-tinted woolly bracts make interesting low verticals about 60–75cm/24–30in high, which thrust up from among a pale carpet of *Artemisia stelleriana*. *Crepis incana* has been in flower for weeks, palest pink dandelion-like flowers with cream centres. There are still buds to come from among the low clusters of deeply notched narrow leaves. *Osteospermum jucundum* continues to produce its bold mauvish pink daisy flowers, while *O.* 'Lady Leitrim' sprawls across the gravel path, chalk-white daisies stained red on the back, with navy-blue centres.

Euphorbia seguieriana is of great value in high summer, introducing that fresh yellow-green when a certain degree of dry tiredness begins to steal into the garden. *Euphorbia stricta* is an annual that produces very dainty heads of tiny acid-green flowers on red-stained branching plants; seeded around the base of *Verbena bonariensis* or among the fresh crop of white daisies on *Anthemis punctata* subsp. *cupaniana* (cut down in June), these accommodating plants add an almost spring-like freshness to high summer.

Where pink saponaria and yellow *Allium moly* reigned in June there is now a richly sombre group: mauve spires of *Origanum laevigatum* 'Hopleys' stand above the dark maroon-tinted foliage of *Euphorbia dulcis* 'Chameleon'. Behind them a grass, *Pennisetum orientale*, moves in the breeze, a haze of furry caterpillar-like flower heads. In total contrast *Sedum telephium* subsp. *ruprechtii* forms a luscious picture – greyish green succulent leaves overlaid with a warm, reddish glow – carry large heads of pink-tinted buds opening into wide, flat heads of yellowish green flower clusters. The combination of colour tones and wax-like texture is beautiful, from its first appearances in spring and throughout the season.

CLEMATIS, AGAINST THE ODDS

Strange as it may seem there are clematis in my Gravel Garden without a wall or archway in sight. I first saw clematis grown horizontally in the garden of the late John Treasure, at Burford House Gardens in Worcestershire. There he was trailing them over a long bank of heathers which normally would have looked very dull late in the year.

When I began the Gravel Garden seven years ago I planted at the same time a 200m/656ft long 'shrubbery' of mixed trees, shrubs and low cover plants to screen the car park from the north-east winds and flat view of neighbouring farm crops. I also planted, optimistically, in the same beach-like mixture of sand and gravel, several clematis, hoping they would survive unpromising conditions and surprise visitors as they drove by them to park their cars on the mown-grass car park. For several years I watered them during the dry months, a 12 litre/2½ gallon can to each plant, which had been heavily mulched with straw. Some died of course, but there were successes. You may pause as you leave your car to look at them, and then later meet some of them again in the Gravel Garden.

In a very poor corner of the Gravel Garden we have trained a *Clematis tibetana* subsp. *vernayi* like a lace shawl over the large leaves of *Bergenia* 'Rosi Klose'. When grown from seed, this clematis varies in size, shade of yellow and thickness of petals, the thickest petals giving rise to the name 'Lemon Peel' clematis. Deceptively dainty as they may appear, I can see I shall have to be ruthless with invading clematis seedlings blown in by the wind.

In the slightly better soil bordering the drive, partially shaded here and there by two ancient oaks, I have experimented with a few clematis to see if they can survive, even thrive, and bring added interest at this in-between season in the garden, twining through shrubs or spreading like a piece of embroidery over quiet mats of periwinkle, *Geranium macrorrhizum* – and, inevitably, over bergenias.

Clematis × *triternata* 'Rubromarginata' is believed to be a natural hybrid between *C. flammula* and *C. viticella*. Despite poor soil (but a good send-off in a really deep hole generously supplied with well-rotted compost) it has flourished, now spreading for yards through a pink-flowered hebe where its multitudes of plum-purple starry flowers with pale centres will take the place of the hebe flowers now beginning to fade. Mingled with it I see *C.* 'Kermesina', with much larger flowers.

I planted *C.* 'Prince Charles' with its soft-fading blue petals and deeper-toned sunken veins, frilly at the edges, among the yellow-leafed *Salvia officinalis* 'Icterina' and grey-leafed *Caryopteris* × *clandonensis*. It can be seen through a screen of *Verbena bonariensis*, echoing its colour, but needs feeding up a bit – or more time perhaps to get its roots down into the bottomless pit of gravel which lies beneath the Gravel Garden.

The four broad violet petals of *Clematis* × *durandii* are still showing among the greyish white leaves of *Brachyglottis* (Dunedin Group) 'Sunshine'. I stole that idea from a picture in Christopher Lloyd's book on clematis, and am surprised and thrilled each year as I see this low-growing clematis clambering back to life, and flowering over a long period, silky green seed heads now forming from earlier blooms.

Clematis 'Kermesina' is pretending to be a cistus. It is growing through *Cistus* 'Peggy Sammons', whose main display was in June, but still lays claim to the bush with a few clusters of pink, crumpled wild rose-like flowers paling towards the centres, carried on long shoots clothed in felted-grey leaves. Fortunately the bushes (two together) seem stout enough to carry the garlands of plum-velvet clematis flowers, appearing paler when backlit by low evening sun. Masking the leggy stems of the cistus are several low bushes of *Ballota pseudodictamnus*. Pruned of their flowering stems in June, they have now refurbished themselves, forming large, tidy mounds of pale grey, felted foliage. *Lychnis coronaria*, the bright, wine-coloured form with almost white-felted leaves, picks up in lighter tone the colour of the clematis.

I would not have expected these clematis to do as well as they have in the situation, but by giving them a good start, waiting patiently and possibly having a bit of luck, they appear to have made themselves good root systems deep down to provide us with unexpected pleasures at this awkward time of year.

THE LAST WEEKS OF SUMMER

PREVIOUS SPREAD If I could choose only one grass I think it might be *Stipa gigantea*. From tidy clumps of green rolled leaves rise many tall stems, held firmly erect, supporting superb heads of oat-like flowers which shimmer as if made of beaten gold. The flowers mature by late June but the overall effect of captured sunlight persists well into autumn.

As summer comes to an end, there is still colour, much of it subdued with dark patches, various purple-leafed sedums in seed, making shadows among the grey foliage plants and gently moving grasses.

We had 50mm/2in of rain during July and August. It fell in dribs and drabs, rarely as much as 12mm/½in at a time, just enough to wash off the dust. This is very little, combined with 'soil' so stony it drains like a colander, especially when you think that many plants have produced their optimum growth by early July and are transpiring heavily. The saving grace this year has been an overcast, miserable sort of summer for most people, while for us (gardeners) it has been a relief not to watch plants roasted by unaccustomed high temperatures. In recent summers, affected I think by global warming, July and August have been anxious months when hot winds from the Sahara 'painted' parked cars with red dust and I felt the blast like the opening of an oven door as I stood on the heat-reflecting stony paths in the Gravel Garden.

During the summer of 1995, when temperatures ranged from 25–30°C/80–92°F and the temptation to water was so intense, some plants showed signs of stress but most recovered and put on surprising new growth as autumn beckoned. By the end of August that year, the almost too lush effect of strong growth and abundant flowers had given way to the kind of garden I had been anticipating. The shining river of gravel became the main feature, the focal point, setting off with its bright warm colour the more subdued tones of the surrounding borders. Subdued, but not dull. Despite the arid conditions still healthy foliage provided colour, shape, textures and form.

Some plants suffered but few died. *Verbena bonariensis* surprised me. 'So good, in 1994,' I wrote, but in 1995 it was half the height and the flower buds withered. To cut our losses we pruned the plants down halfway, long before they were due for tidying up, and by the end of August, satisfyingly, there was a lot of new growth, despite little rain. (How much of that could have penetrated to the roots, I wonder?) The drop in temperature helped, as did cooler nights in autumn with dew trapped beneath the gravel mulch.

Initially the bergenias flopped their large leaves like sad-eared spaniels, but as they adjusted to the heat they stood erect again and, with a 5°C/10°F drop in temperature from the upper 20s°C/80s°F, they resumed their role as handsome definition on the curving border edges, shining rich green before the onset of cold nights when they flush into many shades of red.

Most summers, after several weeks of high-summer drought, typical of East Anglia, there is, each year, a dry look to this area, but not a dead look. Seed heads of many shrubs and plants, including alliums and grasses, come in various shades of fawn and brown, repeating the colours in the pebbles of the paths and making light-catching verticals above many shades of grey and green foliage plants. I am often thankful to see how many plants not only survive but also look good after this testing period of drought and was so astonished in 1995 that I made a list.

The Gravel Garden, 19 August 1995, after eight weeks with no measurable rain, temperatures 25–30°C/80–92°F, plus wind:

Plants looking good still:

Agapanthus hybrids
Allium senescens subsp.
  glaucum
Calamintha nepeta
Euphorbia pithyusa
Euphorbia seguieriana subsp.
  niciciana
Ferula communis
Gaura lindheimeri
Gypsophila ‘Rosenschleier’
Haplopappus glutinosus
Incarvillea arguta
Limonium platyphyllum
Linaria dalmatica
Nepeta ‘Six Hills Giant’
Perovskia ‘Blue Spire’
Saponaria × *lempergii* ‘Max
  Frei’
Sedum aizoon
  ‘Euphorbioides’
Sedum ‘Bertram Anderson’
Sedum ‘Matrona’
Sedum populifolium
Sedum spectabile – all cvs
Sedum telephium subsp.
  maximum – all cvs
Sedum telephium subsp.
  ruprechtii
Sedum ‘Vera Jameson’
Tulbaghia violacea
Verbena ‘La France’
Verbena rigida
Veronica spicata subsp.
  incana ‘Mrs Underwood’

Creepers:

Artemisia glacialis
Diascia barberae ‘Blackthorn
  Apricot’
Diascia vigilis
Phyla nodiflora
Stachys byzantina and *S. b.*
  ‘Primrose Heron’
Thymus ‘Doone Valley’
Thymus longicaulis
Thymus serpyllum ‘Minor’

Retaining good foliage:

Artemisia abrotanum
Artemisia absinthium
  ‘Lambrook Silver’
Ballota pseudodictamnus
Brachyglottis compacta
Caryopteris × *clandonensis*
  cvs
Eriophyllum lanatum
Eryngium eburneum
Euphorbia characias subsp.
  wulfenii
Euphorbia epithymoides
  ‘Major’
Euphorbia myrsinites
Genista hispanica
Genista lydia
Helianthemum cvs
Lavandula cvs
Ruta graveolens
Salvia lavandulifolia
Verbascum bombyciferum
  (young plants)
Yucca

Good seed heads:

Allium cristophii
Phlomis tuberosa ‘Amazone’

Grasses:

Calamagrostis × *acutiflora*
  ‘Overdam’
Poa labillardierei
Stipa gigantean
Stipa splendens
Stipa tenuissima

Affected by drought:

Anthemis
Bergenia – some not all
Cistus
Euphorbia dulcis
  ‘Chameleon’
Hibiscus
Kniphofia
Ligustrum
Malus (later removed)
Sorbus (later removed)
Verbena bonariensis

OPPOSITE TOP Comfortable
companions in dry conditions
include cistus, genista, *Parahebe
perfoliata*, *Gladiolus communis*
subsp. *byzantinus*, *Lychnis
coronaria* and *Salvia nemorosa*
subsp. *tesquicola*, while *Acaena
saccaticupula* 'Blue Haze'
embroiders the gravel with
tiny, pewter-grey leaves and
purple stems. An American
agave, bedded out for summer,
provides sharp contrast.

OPPOSITE BOTTOM
I sometimes find the promise
of flowers still to come almost
more appealing than the full
glory of maturity. *Eryngium
giganteum* fascinates me at
this stage as I like to look
down into intricate patterns
of silvered veins and prickles.
Stipa tenuissima prepares to
flaunt silvery green plumes while
Anthemis tinctoria 'E.C. Buxton'
keeps its eyes shut till later.

A HAZE OF MAUVES AND BLUES

The overall effect in late summer is a haze of mauves, soft purple and blue, with a few yellows that stand out and lift the scene, and lots of silvers and greys all covering the hot soil with apparent relish for their situation. The gaps left when the beautiful sheaves of white daisies on *Anthemis punctata* subsp. *cupaniana* eventually have to be cut down are replaced by new finely cut leaf-rosettes, hugging the stones, their greyness whitening as the dry days continue, but strong enough to produce a late flush of flowers.

In early July the catmint *Nepeta* 'Six Hills Giant' flopped its long flowering stems across all its neighbours, burying I know not what. As it had flowered for weeks I decided to be brutal and cut it fairly hard back. Now it has formed tighter mounds of foliage topped with shorter stems of flower. A tall haze of lavender-blue, almost the exact colour of the catmint, is made by the Russian sage, *Perovskia* 'Blue Spire', which, with the common fennel, *Foeniculum vulgare*, almost in seed now, forms the apex of an interesting group. Filigree grey spires of *Artemisia alba* 'Canescens' tangle among the base of the sage, while entirely fresh growth of *Euphorbia epithymoides* (cut down after flowering) is just coming into welcome late flower. A too seldom used sedum, *S. populifolium*, makes a low bushy plant, smothered now with loose clusters of cream and pink starry flowers, good to pick and very fresh and pretty at this time of year. Golden marjoram is refurbishing itself with bright yellow leaves.

A low group on the curve of a border has stopped me in my tracks for several weeks and still looks good. *Gypsophila* 'Rosa Schönheit' (syn. Pink Beauty) is a low mat-forming plant, whose small grey leaves are buried for weeks beneath single, deep pink flowers. The flush has gone, but enough remains, dotted in small bunches like sequins, to sparkle among the dark shadows of *Sedum* 'Bertram Anderson'. Out of this tapestry emerge fresh flower heads of the late-flowering *Allium senescens* subsp. *glaucum*. It has cool lilac flowers standing barely 30cm/12in tall. This group is completed by the everlastingly beautiful *Euphorbia myrsinites*.

The tall purple *Sedum telephium* subsp. *maximum* 'Atropurpureum' is backed by a spiky yucca, while at their feet, at the gravel edge, sprawl *S.* 'Vera Jameson' and *Artemisia stelleriana* 'Nana'. Upstanding among them, *Tulbaghia violacea* still produces delicate pale lilac flowers, and the bleached grass *Stipa tenuissima* from Mexico and Argentina moves gently its feathery heads above another purple shade, the spurge *Euphorbia dulcis* 'Chameleon', its stained purple leaves now showing signs of autumn with touches of cherry and orange. (As I have said, the Gravel Garden 'soil' is really too poor for this euphorbia but seedlings continue to appear.) For weeks throughout June and July this group was

highlighted with the football-sized skeleton heads of *Allium cristophii*, while *Verbena bonariensis* and *V. rigida* added light and dark purple tints.

SCENE LIFTERS

The light green flower heads of *Euphorbia seguieriana* subsp. *niciciana* have created a focal point wherever they appear, not dimmed by lilies, galtonias or any other star performers which come and go, leaving them still an important part of the stage. The pale moth-like flowers of *Gaura lindheimeri* look well behind – with a great froth of soft brown in front made by the dry heads of *Gypsophila* 'Rosenschleier' (syn. *G.* 'Rosy Veil'). This could not be called eye-catching – downright dull might be more accurate – except that in context, with the bright euphorbia, blue caryopteris and tall spires of Russian sage, its shape and texture look good against the pale gravel. One patch I cut back earlier, exposing the gravel mulch, so new shoots are forming, and it may flower again before the frosts come, usually in November, but sometimes earlier.

Forming a dramatic frieze against the background of the newly clipped leylandii hedge are some big fellows: big brown thistle-like heads of cardoon, *Cynara cardunculus*, and more brown seed heads on *Lavatera cachemiriana*, which was crowded earlier with pale wild-rose-like flowers. The grey-blue young foliage of *Eucalyptus gunnii* is interspersed with the 2m/7ft stems of *Stipa gigantea*, shimmering still in the sunlight, and globular heads of the wild leek, *Allium ampeloprasum*, its silvery pink flowers over now, but its grey-green seed heads standing out well against the background.

Two forms of privet, *Ligustrum ovalifolium*, the yellow and white variegated forms, shine in the east-facing tree and shrub border which edges the entrance road, forming a good backcloth as you face the garden with the sun behind you, repeating all the other yellows which stand out and lift the scene. A few tall verbascums, the leaves of *Salvia officinalis* 'Icterina', almost brassy bright now, the strong vertical lines of the yellow and green variegated *Agave americana*, the softer, fresher, yellow-green foliage of *Caryopteris* × *clandonensis* 'Worcester Gold' – such plants I am grateful for in an area where bedding out to provide colour would not be practical.

The kniphofias are a bit between seasons. There are still a few creamy yellow flowers on *Kniphofia* 'Little Maid', while *K.* 'Nobilis' stands tall with large clubs of orange-red flowers against the leylandii hedge, but planted a good 3m/10ft from the base. Over the past seven years only the stoutest kniphofias have survived several severe droughts: they include *K.* 'Nobilis' and *K. caulescens*.

OPPOSITE *Sedum* 'Herbstfreude' and *S.* 'Ruby Glow' bring a glow of welcome colour to enhance the background furnishing of cistus, ballota, salvia and euphorbia, whose contrasting foliage creates a satisfying design when no flowers bloom. The late-flowering annual *Cosmos bipinnatus* 'Purity' floats its shallow saucers endlessly until caught by the first frost.

GREY AND SILVER PLANTS

Silvers and greys look their whitest and best at this time of year as they have produced new leaf after good grooming in late June/early July after flowering. Most flower in spring and early summer, after which we remove their old flowering shoots and prune to a shapely framework, which is then stimulated to refurnish itself with bright, fresh foliage so that by September most will have taken up their allotted space.

Although I have chosen drought-tolerant plants for the Gravel Garden I am still amazed – and grateful – for their ability to survive and even thrive in the kind of conditions I am obliged to offer. But it is these conditions – long, hot, dry weeks with only odd scraps of moisture, followed by cooler conditions in September – which provoke the grey and silver plants to produce the best foliage by the end of summer, with much of it lasting throughout winter.

Grey and silver plants wear shirts of silk, wool and sometimes wax to protect their thin, green leaves from sunburn, and to reduce transpiration. Some grey plants revert to green when the need to conserve moisture is no longer there, which indicates that plants adapted by nature to dry conditions will grow out of character if planted in areas of high rainfall.

Much as I love grey foliage plants, I think they need varied companions for contrast. Too many greys, herded together, can resemble an ash heap. *Ballota pseudodictamnus* is a star performer among greys. From a woody base spring long, curving stems clothed in round, felted leaves, every tip forming a pale rosette, pretty as a flower, especially so when a well-grown plant presents close-packed rosettes. In full sun and dry conditions the whole plant becomes heavily felted, appearing almost white in colour. The flowers themselves are minute scraps of mauve hidden in conspicuous pale green bobbles carried beneath every pair of leaves. I enjoy the look of these. I always find it hard to cut off these flowering stems, but by mid-June the woody framework is falling apart with the weight, and showing a mass of young shoots needing air and space to grow, so they are pruned to a framework and in a few weeks have reclothed themselves, and will make strong accents throughout the winter.

Helianthemums are especially rewarding, particularly if you have carefully pruned out all the dead twiggy pieces from underneath. Just now, established plants of the form *H.* 'Rhodanthe Carneum' (syn. *H.* 'Wisley Pink') have made flowing mounds of small, oval leaves looking as if they had been stamped out of pale grey felt. Behind is a shrubby plant called *Helichrysum hypoleucum*, with whitened stems and foliage setting off flat heads of tiny yellow flowers that last well into early autumn. My plant has made new growth carrying fresh, green, poplar-shaped leaves, with no need now of the coating of fine hairs which

whitened it all summer. In my drained soil it grows tough and wiry, but it would not survive hard or prolonged frost.

Another plant which takes off its shirt in cooler, damper weather is *Anthemis punctata* subsp. *cupaniana*. In May its low sprawling mounds of finely cut silvery grey foliage are crowded with chalk-white daisies, but in the high summer months when there is scarcely any moisture it struggles to maintain much-reduced, tightly compressed rosettes of ash-grey leaves, yet it never gives up. By early October there will be broad sweeps of bright green, healthy-looking rosettes of leaves – so different in effect you might not think it the same plant, but there is no doubt at all that we shall see once more a meadow-like effect of daisies next year when the plants will have resumed their silvery grey shirts. In the stony places between I see already seedlings of *Euphorbia stricta* whose frothy heads of tiny lime-green flowers will look so fresh and spring-like among the daisies, while another annual, *Omphalodes linifolia,* is coming up as well, its tiny, grey seed leaves easily recognizable.

The santolinas, both grey- and green-leafed forms, look rather elderly. They responded quite well to the July trim, even, I can see on close examination, breaking new shoots from their furrowed woody trunks. It seems heartless to throw them out when they are making such an effort. In any case I shall wait until early spring to decide whether or not to replace them with young ones. Sometimes it is best to be ruthless.

I am frequently asked what to do with coloured-leafed forms of *Salvia officinalis*, the purple or yellow variegated sages. They do become very straggly after a few years. Much depends, I think, on the conditions. Too much wet does not suit them. In reasonable soil where you have plenty of fresh young growth it is possible to give them a trim in early spring, reducing their spread if that is required. Even if they look a bit sad and bedraggled by the end of winter, they refurbish themselves, making large piles of beautiful velvet-textured leaves. But on my arid gravel they tend to make, after several years, a bare twisted framework, with new growth all at the ends of skinny branches. Yet again, looking closely, I can see tiny new leaves bursting from bare wood. They are triers. However, young plants, pot grown, grow away so quickly, making such an attractive effect with larger, better-coloured foliage, that I feel certain we will replace the six-year-old bushes in the coming spring, refreshing the exhausted soil with a bucketful of compost to give them a good start.

One of the most striking greys just now is the old-fashioned lambs' ears, the non-flowering form, *Stachys byzantina* 'Silver Carpet'. It thrives in poor conditions and has spread its velvety leaves for yards where most plants would have packed up years ago. The flowering form *S. b.* 'Cotton Boll' we named because its minute flowers are held in round bobbles at the top of stiffly upright stems, so heavily buried in white wool they reminded me of the fields of cotton I saw growing in North Carolina. We cut bunches of them when they are at their whitest, to dry for dried flower arrangements. *Stachys byzantina* 'Primrose Heron' becomes anonymous throughout the dry months, as grey as the rest, but now its leaves are already showing lemon tints and will become brighter throughout the winter months, well into spring. *Stachys byzantina* 'Big Ears' has larger leaf rosettes than any of the others, less densely coated, so it appears grey rather than silver, but I like its bold, healthy look and value its ability to make a wretched situation look positively easy.

David Ward has just passed by as I sit on my little stool writing, to say that in his garden sparrows decimate his grey plants for nest-building material. I have never noticed that here. Perhaps the balance of the birds to foliage is right for the plants! But I have noticed sparrows doing good – catching white butterflies to feed their young – and David confirms this, saying they go into our polytunnels to chase and catch the white butterflies.

ADDING IMPACT

Much-needed accents among all the soft, woolly features are provided by several different yuccas. Their grand impact does not appear overnight. It has taken a few years for each plant of *Yucca recurvifolia* to make several crowns and now form a strong group of dark grey-green, sword-like leaves in the centre of one of my island beds beneath the airy fountain-like shape of a young Mount Etna broom, *Genista aetnensis*. In another island bed, two different yuccas flank the tall pencil-shaped *Juniperus scopulorum* 'Skyrocket'. One, the slow-growing *Yucca gloriosa* 'Variegata', is sending up at this late stage its stout stem of beautiful flower buds, each individual flower enclosed in one long, pointed, mahogany sheath. It will barely have time to release its ivory bells before the frosts arrive – and, worse still, the crown of handsome yellow leaves which produce the flower stem will die after flowering. But I am thankful to see three more young crowns thrusting up from the base. (Another young plant elsewhere is also producing a flower stem, but I see no new crowns there. What will happen, I wonder? I shall leave it, hoping something may appear next season.) The other yucca might be a form of *Yucca flaccida*. It is not *Y. gloriosa*, which I have elsewhere in the garden and is much larger, with elephantine trunks carrying great crowns of leaves up to 2m/7ft tall with broad, stiffly pleated leaves. This plant has not attempted to make such thick woody trunks, but has a dense stand of many crowns at ground level. Later, it makes a splendid background for airy screens of *Verbena bonariensis*, warm chocolate-brown leaves and seed heads of *Sedum telephium* subsp. *maximum*

PREVIOUS SPREAD In late
summer this lazy stemmy look
replaces the lush growth and
waves of colour seen earlier
in the year. The gravel soil is
parched, the air dust-dry. Only
the toughest remain – seed
heads, grasses and deep-rooted
or tuberous-rooted perennials
such as the *Nepeta tuberosa*
in the foreground. After an
exceptionally hot spell, an elderly
botanist was heard to say, 'This
looks – and smells – just like
the Mediterranean.'

'Atropurpureum', and the pale mauve starry-eyed umbels of tulbaghia, which flower on and off for months and look pretty intermingled with verbena.

Two large bulky plants create a grand effect opposite one another across the pale gravel path, which itself is as much a focal point now in the muted landscape as any of the plants. One is the silver-leafed bush, *Atriplex halimus*, which has been trimmed several times throughout the growing season and is crowded now with such pale lacquered leaves it stands out conspicuously as a shimmering mass, easily 1.5m/5ft tall and as much across. Opposite is *Melianthus major*, a sumptuous plant for the end of the season, it never fails to make me catch my breath. It is, when well grown, one of the most beautiful foliage plants we can grow in the garden. Sitting on my little fishing stool, I am looking at a mature clump almost 1.5m/5ft tall and much more across, with possibly fifty stems each clothed in large pinnate leaves with sharp-toothed edges, almost blue-green when mature, pale pea-green when they unfold from beautiful long pale bracts which clasp the stout stems. Each leaf arches over as it expands, as if to make way for the young ones coming after, so the whole plant is a symphony of curves, of light and shadows, and highlighted, as today, with glistening drops of water held like diamonds along the veins. This plant comes from South Africa so is not reliably hardy in cold soils or districts, but it has survived here for many years, in well-drained gravel, and must have a very deep root system to be able to make such a display after months with very low rainfall. In recent mild winters there have not been enough frosts to damage the foliage, but I like to cut the old stems to the ground in spring, to encourage fresh, clean growth. If left it will produce dull-looking, club-like heads of small chocolate flowers which I can well do without, accompanied as they are, by spring, with tatty-looking overwintered foliage. The space left after cutting down is carpeted with spring bulbs using the top layers of soil, unharmed by the melianthus roots living far beneath them.

PLEASING PARTNERSHIPS

Where the soil is slightly better, alongside the entrance drive to my house, where the site is partially shaded during the afternoon by an ancient oak, I have been enjoying the partnership of a rose and a clematis. The rose is a collected form of the wild Burnet Rose, *Rosa spinosissima*. On reading *Shrub Roses of Today* by that great authority Graham Stuart Thomas, I think it must be *R spinosissima* 'Falkland', according to his description. Its great advantage for me is its tolerance of poor, sandy soils and its ability to make a densely twiggy framework capable of supporting a scrambler. A disadvantage might be its suckering habit, but with something so attractive and amenable I don't find that a problem. A quick chop round the main clump with a spade in early spring easily dislodges the wanderers, which can be potted and quickly established for eager visitors who see the bush in flower in early June and can't bear to leave without one. They are enchanted by the small, incurved semi-double roses, pale pink with a touch of lilac, seen against a background of tiny grey-green leaflets which furnish the bush so completely you cannot see through it. By September the roses have long gone, but if you look carefully you will see blackcurrant-like hips, equally round, black and shining. In my case they are partially smothered by *Clematis* 'Étoile Violette'. This clematis returns each year from underground shoots, thrusting up through the rose over which it is now draped like a bedspread laid out to dry, embroidered all over with deep velvety-violet flowers. Trailing stems fall to the ground, mingling with the pale grey filigree foliage of *Artemisia absinthium* 'Lambrook Mist' (an improved selection of *A. a.* 'Lambrook Silver' I imagine), probably at its best now, like so many grey plants in late summer and autumn. (Since writing of this partnership we have given the clematis a tall tripod of home-grown bamboo canes, to the relief of the burnet rose.)

Nudging into the large, silvered, feathery shape of the artemisia are fresh-grown plants of *Euphorbia epithymoides*. It was hard to cut them down in late June, after we had collected seed, but despite dry weather in July and August they have made plenty of new growth, with no sign of mildew, and are already showing a fresh crop of yellowy green flowers. These repeat the yellow foliage of *Origanum* 'Norton Gold', which edges the gravel path, where trailing stems of clematis – both flowers and silky green seed heads – make a rich pattern among it at ground level. The Russian sage, *Perovskia* 'Blue Spire', with its whitened stems and long, branching spires of lavender-blue flowers, provides a delicate vertical screen, a contrast in shape, but repeating, in softer tone, the violet clematis.

A new addition, planted only this spring, provides a sweep of contrasting colour towards the front of this group. It is a newly introduced gaura, *G. lindheimeri* 'Siskiyou Pink'. So far it appears to be shorter than the white form, about 45cm/18in tall, but several plants grouped in a curve have made up surprisingly well, spreading a wash of rose-pink into the foreground. Wire-thin stems bend in the breeze, bowed with spires of four-petalled flowers, almost white at the edges, deepening to rose in the centre. Despite their overall fragility, the massed effect, of hovering rose-pink butterflies, is a joy.

Framing this large mixed group a bold curve of massed *Bergenia* 'Mrs Crawford' is just beginning to assume warm autumn tints, while beside the gaura are soft green and beige feathery flower heads of the grass *Stipa calamagrostis*.

At the back of this border, along the driveway leading to the house, the planting is simple, composed of fewer species, each covering more space. The wide curve of bergenia sweeps round the back of *Rosa spinosissima* 'Falkland', till it meets the huge specimen of *Helleborus × sternii* 'Boughton Beauty' which has as many as fifty stems already falling out from the central point. Low afternoon sunlight accentuates the purplish-red colour of main stems and leaf stalks, while the dark, claw-shaped, leathery leaf clusters will protect the terminal flower buds throughout winter. Yards of *Vinca minor* 'La Grave' (a dwarf free-flowering periwinkle) and the aromatic cranesbill *Geranium macrorrhizum* 'Album' have provided both ground cover and support for a variety of colchicums. Most of them are past their best now at the end of September, but there are still a few to come to take us into October. Colchicums grow wild in high mountain meadows, where herdsmen from the valleys below take their cows to graze in summer, creating short turf through which colchicum flowers emerge in autumn. Although generally known as autumn crocuses, they are not crocuses at all. They are now considered to belong to their own family, *Colchicaceae*, while crocuses belong to the iris family. See the plan on page 18–21 for the planting of this island bed.

PLEASURES STILL TO COME

The larger sedums are still to come – or more accurately have been coming for weeks, ever since the late summer, especially forms of *S. spectabile.* I love every stage in the development of these flowers. Initially their large flat or domed heads of waxen buds are pale grey-green, changing to resemble pin-head oatmeal with touches of pink breaking through until their fully open flowers warm the autumn garden with shades of pink, purple or brick-red. Already honey bees are foraging for opening flowers on *Sedum* (Herbstfreude Group) 'Herbstfreude', while near by the blue flowers of *Caryopteris × clandonensis* 'Arthur Simmonds' are vibrant with the humming sound. Throughout the Gravel Garden the air is alive with insects sipping this scented harvest, concentrated by the sun.

A plant new to me is *Salvia* 'Indigo Spires'. For weeks it has shown no signs of stress, its square stems clothed in dark matt leaves, topped with long, thick tapers of closely set, almost navy-blue calyces, from which pout vivid violet-blue lipped flowers. These dark threaded calyces form a feature long after the flowers have gone lower down, but there are terminal tips of buds waiting to open, promising yet more pleasure to come. With more experience I would now consider this plant to be tender. This single specimen is set off by the interesting bronze and grey-green foliage of *Euphorbia epithymoides* 'Candy' with a carpet of sweet lemon-scented *Thymus herba barona* creeping into the path at its feet. Beyond, upright new stems of *Cistus* 'Peggy Sammons' make a suitable background, repeating the vertical lines. But to my surprise again, some of the cistus scorched in the days of greatest heat in 1995. I think because of the good feed of compost initially, combined with an extra wet winter, the plants made excessive and unnatural (for them) growth. None died, but in some cases branches died back. We pruned carefully, removing stressed branches and shortening others, to reduce the size overall a little and to help reduce strain if and when it comes again. In the case of *Cistus* 'Peggy Sammons' it is easy to cut out some of the tallest and oldest pieces where there is plenty of young growth adapted to tough conditions.

Another flat ground cover which does its job very efficiently and attractively is *Leptinella potentillina*, much more familiar to me as *Cotula potentilloides*. It doesn't turn a hair during dry times, pushing out a flat carpet of green and bronze small fern-shaped leaves flat into the gravel. Out of this flat plain stand good verticals. First the smallish *Diplarrena moraea* with neat upstanding rushy leaves which held earlier, in midsummer, exquisitely formed three-petalled white flowers, while now the green-flowered *Galtonia viridiflora*, with much larger strap-shaped leaves and 75cm–1m/30in–3½ft stems, dangles pale green bells above a huge rosette of the largest leaf bergenia, *B.* 'Ballawley'. All greens together, but to me very satisfying. Just beyond, colour is coming in the heads of *Sedum spectabile* 'Brilliant' picked up at ground level by the soft mauve flower heads smothering the green and yellow variegated thyme, *Thymus* 'Doone Valley'. There is more promise of pleasure to come with *Caryopteris* × *clandonensis* 'Heavenly Blue', the darkest blue form, not yet showing colour, while the flower heads of the Japanese grass *Miscanthus sinensis* 'Yaku-jima' are emerging at shoulder height and will carry us into winter when the entire plant, bleached to pale straw, will catch the low sun sinking in the west.

DEPENDABLE SEDUMS

Modest plants can mean as much to me as their flamboyant neighbours. Too many star performers lumped together can become quarrelsome; neutral areas between them keep the peace. Sedums make very good peacemakers and are so dependable on hungry soils, providing interest for almost every month of the year; few herbaceous plants can do that. The majority of plants, especially those with glamorous flowers, seldom last as long as a month, often much less. I am thinking of much-loved beauties like border iris, lilies or exotic oriental poppies. On a still June morning we draw in our breath, marvelling at the silk or velvet-textured petals, the haunting beauty of rainbow colours, but all too soon the season has ended – far too soon if wind and rain assault them.

Gardeners in heavy wet soils might find sedums liable to rot, or grow lanky and leggy since these plants are perfectly adapted to dry conditions. They have fat succulent leaves which store water, while many are coated with a protective waxy bloom. I am grateful to the many different sedums which thrive in sand and gravel soil, and survive drought, which is a way of life here, not an occasional hazard. I possibly should admit here that these conditions are probably the limit of what the larger herbaceous sedums will take. Prolonged scorching temperatures combined with scanty rainfall and no irrigation might well be too much for them. But so far, after seven years of typical Essex summers, they seem well adapted and make compact, free-flowering plants, while elsewhere in the garden, in richer, damper soil, they grow taller, with heavier heads, and so tend to fall out untidily from the crown centres.

Planted in bold groups (one alone can be effective), their seed heads add impressive deep tones among all the muted shades of winter, furnishing the area until we cut them down in February to make way for the new shoots emerging at ground level. From March onwards every stage of the developing plants is attractive, since they quickly form large tidy clumps of foliage, jade-like in colour and texture, of an icy smoothness which makes fine contrast with woolly lambs' ears, *Stachys byzantina* 'Silver Carpet' and felted *Ballota pseudodictamnus*, while spring bulbs, scillas and chionodoxas form pools of blue, self-sown between them. By midsummer I value the large dome-shaped heads of jade-green flower buds as much as when their opened flowers flood the space with colour – crowded platters of tiny star-shaped flowers in shades of rosy-mauve in *Sedum spectabile* 'Meteor' and 'Brilliant', while 'Iceberg' is a cool white. All are humming with bees in late summer and early autumn.

The well-known *S.* Herbstfreude Group 'Herbstfreude', a hybrid between *S. spectabile* and *S. telephium*, attracts bees as well as visitors with brick-red flowers, with no hint of mauve for those who find this shade disagreeable.

TOP *Sedum* (Herbstfreude Group) 'Herbstfreude' in winter

BOTTOM *Sedum aizoon* 'Euphorbioides'

Sedum 'Matrona' I brought home from a nursery in Freiburg, where the owner, Ewald Hugin, a first-class young plantsman, selected and named it from a batch of his seedlings. Dark upstanding stems, 60–76cm/24–30in and more, are clothed in fleshy leaves, flushed purple and grey. In August they carry wide flat heads of pale rose-pink flowers, a beautiful combination captivating most visitors to the garden. In late autumn and winter, after the leaves have dropped, I value the dark seed heads for their effect in a muted colour scheme. The seed heads of these sedums vary according to the flower colour, hence *S. spectabile* 'Iceberg' has beige heads while others vary from warm caramel to dark coffee grounds.

Sedum 'Joyce Henderson' is very similar in appearance with slightly more domed flower heads, but it tends to fall outwards from the centre while *S.* 'Matrona' remains firmly upright. (I have just remembered *S.* 'Joyce Henderson' is planted in richer soil and maybe should move her to a more hungry soil to judge her fairly.) I have other forms of *S.* 'Joyce Henderson', obtainable from different sources. Obviously plants grown from seed will show slight variations. I sometimes feel that offspring of any species with only slight variation, introduced as 'new' named forms, are merely confusing – but we each have the freedom to choose which we prefer and ignore the rest!

Another recent introduction has been *S. telephium* subsp. *ruprechtii*, a superb plant for full sun and any soil but soggy. It is a smaller, denser plant than *S.* 'Matrona', but with similar pewter-coloured foliage. Flowering a fortnight or so earlier, it opens surprisingly creamy yellow flowers from pink buds, which mature into reddish brown seed heads. The whole plant fascinates me from spring until autumn.

As I have said, *S. populifolium* from Siberia is often overlooked, unfairly I think. Unlike any other sedum I know, it makes a small bushy plant of woody, much branched stems, covered with light green poplar-shaped leaves with deeply cut edges. By late summer these are swamped with hawthorn-scented pink and white flowers.

The dark chocolate-leafed sedums are important as contrast among grey and silver plants. *Sedum telephium* subsp. *maximum* 'Atropurpurem', with stout dark stems carrying thick fleshy leaves of bloom-coated purple-brown – the colour of chocolate – makes a dramatic feature plant from spring until autumn, but it does need well-drained soil to avoid rotting. In late summer flat heads packed with small starry flowers combine rosy pink and brown. For an artist recently, who wanted me to pick her a bunch of autumn colour to paint as still life, I added the blackish brown stems and seed heads of 'Atropurpureum'. Looking into the seemingly dull flat heads she was delighted to see a warm range of shades that made up something neither brown nor black, but gave her the strength she needed in her painting. *Sedum telephium* 'Möhrchen' is an outstanding variation of *S. t.* subsp. *maximum* 'Atropurpureum', making shorter, more compact plants clothed in succulent dark chocolate leaves, almost smothered in August with dense heads of dark buds opening creamy pink flowers, which fade to deep coral pink, a lovely contrast to dark foliage.

An old variety which has only recently come my way is *Sedum* 'Munstead Red'. Purple-stained stems, clothed in blue-grey leaves, carry wide flat heads of raspberry-red flowers, opening in mid-August. Quantities of new shoots bearing unopened buds appear from leaf axils all along the stems, from ground level upwards, to make a continuous show well into autumn. By late September to October, the plants look like glowing coals, with many shades of red, from bright tones of freshly opened flowers to deep bronze of mature heads.

By late summer some of these herbaceous sedums may be festooned with coils of a cobwebby material forming protected nurseries for crowds of tiny caterpillars packed inside devouring every leaf if we do not catch them in time. An evening visit by a party of moth enthusiasts revealed (to their delight) that these are the larvae of a moth very uncommon in Essex, called *Yponomeuta sedella*. Apparently these larvae were described as common in the Witham area in 1860, and another record came from Colchester in 1908. It was assumed these were feeding on *Sedum telephium*. No further records came until 1985, when a single moth turned up at Saffron Walden; then another was found at Colchester in 1989. I am obliged to Brian Goodey of Colchester for this information. He goes on to say this moth could possibly be found to be widespread in Essex if searched for on garden plants, so not to feel guilty if we are obliged to spray insecticide to save the plants. We avoid commercial sprays as much as possible in the garden but occasional heavy infestations of trouble are dealt with promptly. We would not wish to exterminate this creature, but some control is needed.

The small sedums, or stonecrops, tend to become lost in the large setting of the Gravel Garden and are planted in the Scree Garden (see pages 172–4).

AUTUMN
TAPESTRY

On autumn and winter evenings, when I drive through my gateway, pausing to stop and close the gate behind me, I never tire of the scene caught in the car lights. So many different shapes and textures welcome me. On still October mornings with gentle sunlight, I sit and dream once more on my fishing stool and am astonished to see quantities of tiny violet-purple berries tucked among the small leaves of *Lonicera pileata*. You would scarcely notice them without stopping to search, but once found they are such a lovely colour; seen with the light passing through them they look like shining glass beads.

Here, I originally planted seven to ten plants of bergenia, about 30cm/12in apart, which have extended into a huge colony. I had to learn when I began planting on a larger than average scale to take into account the foreshortening that occurs with length. In a long border, groups need to be extended; otherwise they will appear like dots from a distance rather than a bold group. Now the bergenias dominate the border edge, as piles of huge water-washed boulders do in some modern garden designs. Their health and vigour, in October, belies the poor soil they are covering. For contrast to their large rosettes of glossy leaves I used the modest lonicera with shining box-like leaves and hidden purple berries. In better soil this shrub grows much taller and bulkier, but here it grows only about 50cm/20in tall.

We used to have *Phlomis russeliana* planted behind and its bobbly seed heads lifted this group all winter, but recent hot and dry summers were too much for it. We substituted *Artemisia pontica*, whose filigree foliage makes a base all summer and autumn for the beautiful columnar grass *Calamagrostis × acutiflora* 'Karl Foerster'. The apex of this group is formed by *Juniperus scopulorum* 'Skyrocket', while leathery green cistus, felted *Brachyglottis* (Dunedin Group) 'Sunshine' (syn. *Senecio* 'Sunshine') and ubiquitous *Euphorbia characias* subsp. w*ulfenii* fill in plenty of background.

As I look along my dried-up river bed, sunlight is picking up autumn tints, warming the overall muted tones of overwintering plants. In the foreground, a fine established group of *Agapanthus* 'Isis' makes a glowing sunburst of colour, the strap-shaped leaves now all shades of buttery yellow, contrasting with carpeting plants like helianthemum and star-fish sprawls of *Euphorbia myrsinites*.

Beyond is an island bed bordering the driveway leading to my house. Here the soil is slightly better, not quite so stony, and when the sun is high, in midsummer, and during weeks of drought, some respite is given by two ancient oaks creating partial shade, especially in the afternoon. For some reason the planting here fits so comfortably together, effective both in summer and winter. It is a balanced design, using bulbs, perennials and shrubs thus involving shape, texture and colour, of which enough is retained all the year round to maintain a harmonious whole (see the plan pages 28–9). Yet I feel it would be a mistake to repeat this formula throughout the Gravel Garden. It could be too constrictive, too tidy, too formalized; it would not allow the space and freedom needed for longer stretches of herbaceous perennials which disappear in winter, for large patches of self-seeding annuals, various alliums or spring bulbs. One just has to accept that not everywhere can be furnished with foliage or form throughout the year; that some areas will have their high moments, or even months, and then fade discreetly, while another group, chosen to fit the time of the year, will introduce a new picture.

STARS IN THE FADING LIGHT

Included in the island bed bordering the driveway, the snowy mespilus, *Amelanchier lamarckii*, now dominates the group, turning wondrous shades of red. The leaves on the tip shoots are brightest cherry-red, and gold and amber on the more shaded branches deep in the heart of the shrub. This shrub is a European hybrid of uncertain origin, preferring better conditions than I am offering here, but although I should know better I am no different from most obstinate gardeners: I love this plant, and had to try it. Obviously we gave it a good start, about six years ago, by preparing a deep hole filled with a couple of barrowloads of compost. That will have long since gone, but a strong root system must have developed deep down to support the top growth. And this summer has been kinder for plants than the previous two, with not such high temperatures, and with 134mm/5¼in of rainfall, from April to September. Now I am delighted to measure this year's extension growth on the amelanchier of up to 60cm/24in.

Since repetition helps lead the eye into the distance, I have planted two more amelanchiers in the Gravel Garden. I have to confess the one on the sunniest side is as yet half the size of the other two (they are both about 2.5m/8ft high and across), but it makes its point none the less, a little bonfire shining against a background of pale variegated privet.

Large groups of sedum echo, in darker tones, the colourful amelanchiers. *Sedum* (Herbstfreude Group) 'Herbstfreude' has deepened to dark velvety red, while *S.* 'Matrona' is still magnificent, its dark-stained stems carrying deep reddish brown heads, which glow when a shaft of sunlight touches them yet contribute an effective dark shadow when the light passes, making dramatic contrast with silvery grey *Stachys byzantina* and the still foaming pea-green heads of *Euphorbia seguieriana* subsp. *niciciana*.

Another shrub I have planted for autumn colour is the stag's horn sumach, *Rhus typhina*. I have had it here before, so I know it will take some time to make its mark in these poor conditions. Already I see the difference soil can make. Just 0.8km/½ mile away we planted a tree and shrub border in the village, where in

OPPOSITE I was surprised to find that berberis did not take kindly to low rainfall and hungry soil. *Berberis thunbergii* has been slow to establish but good drainage has resulted in brilliant autumn colour.

one season, they have made twice the growth of mine. But once deep-rooting plants have had time to delve deep into gravel and sand they often surprise us with their ability to survive adverse conditions. Only by experimenting can we discover how far plants will go, what must be their limit.

I was sad to see a fine specimen of *Rhamnus alaternus* 'Argenteovariegata' taken out to make way for the new teahouse we have built at the end of the Gravel Garden to provide a little shelter and simple hospitality for our visitors. Other established plants that had to go were some fine forms of berberis and two lofty conifers, all making a handsome group, bulkily attractive in winter and summer. I had the pleasure of their company for many years, watching them grow and develop as we watch over sons and daughters, but I found courage to part with them, to make new plans for the future. And fortunately I have a fine rhamnus elsewhere. It is not unlike a silver variegated privet at first glance, but distinctly different when you compare the two, both from a distance or with a branch of each in your hand. From afar the privet forms a more open, spreading bush, a pale cream accent between dark yew and small-leafed holly. Its leaves, larger than those of rhamnus, are a light green with irregular cream margins. The rhamnus forms a very upright shape, pear-shaped from the base, as all its lateral branches reach outwards and upwards to display the backs and fronts of smaller, dark-green leaves, marbled with grey, with creamy white margins, the overall effect from a distance being almost silvery, rather than the pale cream effect of privet. Even more distinctive are the dark-stained stems and next year's tiny flower buds clustered in the leaf axils, a warm mahogany-red.

Across the wide stretch of gravel cutting round the front of the house, the now leafless Swedish cut-leafed birch, *Betula pendula* 'Laciniata', forms a filmy fountain of delicate drooping branchlets, screening the glory of a huge *Cotinus coggygria* Rubrifolius Group, which still has every leaf intact, in brilliant shades of scarlet and crimson. It is backed by a tall bulky *Chamaecyparis lawsoniana* 'Ellwoodii', with, inevitably, bergenias and cranesbills swirling round its feet. Seen from my kitchen window this group warms my heart on dull drizzly days and takes my breath away when lit by the rising sun, or again at teatime when bathed in a gold and apricot sunset spreading all along the garden's western boundary.

Clematis tibetana subsp. *vernayi*, its yellow lanterns extinguished now, draped across a wide curving mass of bergenia, sprawls recklessly across the gravel drive, where fallen leaves from the 300-year-old pollarded oak form a brown carpet contrasting with hundreds of seed heads held in the fishing-net tangle of stems. The seed heads can be found in all stages, from fresh green shining silk enclosing the dark ovary, to the mature head, a white wig of dry fluffy awns attached to the central boss of seed cases just waiting for the moment when it all breaks apart

and the wind whisks each viable seed to its new home. One of our autumn jobs is to trim off the mature seed heads before they create a multitude of problems.

I can't help having a soft spot for *Bidens ferulifolia*. In recent years it has become ubiquitous as a bedding plant, especially in hanging baskets where it can look garish with too many competing colours, but is that a good reason for being snobbish about it? Plants are rarely at fault. It is how we use them that makes them good or bad. I cannot allow myself many half-hardies in my general planting since I prefer to rely on a succession of perennials, together with woody plants, but the persistent little bidens, which struggled for weeks unaided in the dry summer months, suddenly leapt away with the least bit of encouragement. Now it makes trails of little starry yellow flowers in unexpected places where David bedded them out after fear of frosts, in empty spaces (which had been a sea of colour earlier) above resting bulbs of scillas, chionodoxas and other spring pleasures.

Gaura lindheimeri looks as good now, in October, as it has since June. Slender stems, much branched, still carry white moth-like flowers, hovering at all levels up and down the stems, the effect warmed by rose-pink buds and calyces. They are amazing plants. Since much-needed rain has come, they have quickly produced fresh growth crowded with spires of new buds. They will continue to open fresh flowers until a sudden sharp frost finishes them for the year.

I cannot overpraise *Euphorbia seguieriana*. Plants placed strategically here and there along the central pathway, together with a few incidentals flaunting themselves in places I would not have considered, still fill me with intense pleasure, their red-stained stems supporting soft masses of brilliant green lace-like clusters of seed heads, bracts and tiny leaves. The plants form such vibrant masses of freshness still, in this autumn setting, no hint of weariness here.

Recently I was invited to the funeral of 'Miss Marple', the actress whom we knew as herself, Mrs Joan Butler, and professionally as Joan Hickson, and sometimes met for little chats when doing the weekly shop in the nearby village. For her I made a simple garland, as I had learnt to do from Hilde Satori, one of my German students. The garden had been lashed by gales the night before, so it was easy to find broken whips of pliable willow, to form into a circle and bind with soft string. On to this base I bound small bunches of this delicate euphorbia, stronger heads of green hydrangea and shining leaves and buds of *Skimmia* × *confusa* 'Kew Green' for contrast. I added rose buds from two roses which flower well into autumn. One, *Rosa* 'Bloomfield Abundance', grows in the mixed shrub border leading to the house alongside the Gravel Garden. For months it produces huge, airy sprays of exquisitely shaped buds, which open small, shell-pink double flowers. The other, *Rosa* 'Pink Grootendorst', has deeper, rose-coloured flowers which make me think of pinks, since each small, double flower has finely fringed petals. Finally I found a few flowers on *R. × odorata* 'Viridiflora', more curious than beautiful, but the strange green and bronze petals (really modified sepals) were perhaps a nod of appreciation for Joan's tartly witty observations.

This digression has led me away from the subject in hand. I must take you back to the Gravel Garden, where, not to be outdone by *Euphorbia seguieriana*, *Gypsophila* 'Rosenschleier' (syn. *G.* 'Rosy Veil') carries a late crop of tiny shell-pink flowers caught in a wiry network of midsummer-flowering stems. They twinkle like stars in the Milky Way.

In recent years many different forms, both species and hybrids, of diascia have been introduced. They come from South Africa, some unreliably hardy, yet so immediately attractive they never fail to tempt us into trying them. A few cuttings in a greenhouse kept just above freezing will ensure good plants for the coming season. The more hardy types are found, I imagine, in higher latitudes which experience some frost. Among them is possibly *Diascia integerrima*, which has never failed here. It makes a display for months, still full of flower, warmed by low evening sun. Sprawling into the gravel path, this diascia is planted in perhaps the worst, most hungry soil of the Gravel Garden. It vanishes completely in winter. Each spring I visit the empty space and wonder if it will return. So far, it always has. The plant slowly colonizes by underground stems (in such poor conditions it has few competitors), making low bushlets of intertangling wiry stems set with small, narrow grey leaves. From early summer onwards it produces tapering heads of small, deep rose-pink flowers. It is so free-flowering that the border edge it furnishes seems to glow against a background of cool, filigree *Artemisia pontica* and the silvery grey foliage of *Achillea* 'Moonshine'. The late summer-flowering *Allium carinatum* subsp. *pulchellum*, with heads of powdery-mauve bells, has seeded meadow-like into open gravel-mulched spaces between all these. During the worst of the dry time the diascia puts up shutters, and sits tight, flowerless till the cooler, damper days of September, when it pushes out a mass of new growth from the dry resting stems, to make, as now, a final flush of heartwarming, curiously spur-shaped flowers.

It is past five o'clock. The sun is sinking fast. Most of the Gravel Garden is dark with shadow. Although my hands are stiff with the cold air dropping suddenly like a shower on to my knees and writing pad, I need to catch for a moment the picture made by the sinking sun as it spotlights the group of a pale columnar grass, *Calamagrostis × acutiflora* 'Karl Foerster'. So quickly the light moves, I can watch, as it dips below the farm headland, the last rays moving upward from the base of needle-like stems till they reach the straw-coloured flower heads outlined against the shadowed garden beyond, till the moment when each column disappears, as the light is switched off.

OPPOSITE TOP Island beds provide the chance to see plants outlined against a void. The seed heads of *Verbascum bombyciferum* and thin stems of *Verbena bonariensis* form an open screen through which to view the entrance drive behind them.

OPPOSITE BOTTOM This group pleases me all winter. The sunburst effect of the grass *Eragrostis curvula* contrasts with the enamelled leaves of *Bergenia* 'Abendglocken'. The pale felted leaves of Jerusalem sage, *Phlomis chrysophylla*, are partnered by the silver-leafed shrub *Atriplex halimus*.

EYE-CATCHING PLANTS

Autumn is a good time to see which plants are making bold statements – full-stop plants I sometimes call them, coming at the end of a sentence. Plants which by their shape, scale and texture cause you to pause and take breath after absorbing what might appear a bewildering combination of small-scale plants with intricate foliage. Boldest of all are the huge basal rosettes of *Verbascum bombyciferum*. They vary in size, but can be more than 1m/3½ft across. With long, felted leaves almost white in contrast with everything else, they lie on the ground, several layers of leaves pinned together in the centre, like a great dahlia head, demanding admiration, both of colour and shape. Such grand yet simple beauty – I almost wish they would never grow up to produce the equally dramatic candelabra of flower in midsummer.

Very different, but also standing out from their neighbours, are the new leafy stems of *Euphorbia characias* subsp. *wulfenii*, standing up to 1.25m/4ft tall and across, which have replaced those we cut down in July, after we had collected seed. They are remarkably handsome in winter, a focal point wherever they happen to be. Mature plants are large and statuesque, with long, narrow leaves arranged around sturdy stems, somewhat resembling huge bottlebrushes, and all a wonderful blue-grey colour, pale where the light catches the backs of upstanding leaves, deep pewter in the shadows. This colouring is enhanced in late autumn by the russet tones of sedum seed heads and scarlet and crimson leaves on several different bergenias.

Astelia chathamica (syn. *A. c.* 'Silver Spear') is an experiment, and may not prove to be hardy; certainly it would not have survived in the kind of winters we had about ten years ago. It is a phormium-like plant, but less aggressive, forming fans of sword-like leaves, pale, shimmering green on the top side, white beneath. In hot, dry weather the whole leaf becomes protected with a silvery patina.

On a smaller scale my eye is caught by a little colony of our native yellow-flowered horn poppy, *Glaucium flavum*, still to be seen on east coast shingle beaches and sand dunes (and probably elsewhere along British shorelines). It forms rosettes of blue-grey pinnate leaves, finely cut and crimped at the edges, and is adapted to dry and windswept conditions by having both a deep rooting system and a dense coating of fine, short hairs which, as I stand writing beside it, are meshed with fine water droplets, the remains of this morning's fog. The blueness is more evident on young leaves at the centres of the rosettes. As the old leaves lengthen, the covering of hairs becomes more widespread, more sparse, and so the leaves appear greener. Established plants form a woody rootstock from which numerous rosettes arise. The poppy stands out here among the little collection of coastal plants, which includes sea thrift, sea campion and sea kale.

OPPOSITE *Sedum* 'Matrona, introduced from Freiburg, Germany, is superb. Here its richly coloured seed heads and stems contrast with *Euphorbia seguieriana*, astonishingly fresh-green after months in flower. Tangled among bergenia leaves is *Bidens ferulifolia*.

The kale's large, wavy leaves have collapsed and been removed, so now the deep roots rest beneath the gravel mulch, as they do along the coastline in winter. The empty space left behind does not disturb me as it might if it were uncovered dark soil since alongside the gap there are the dark green tussocks of sea thrift, contrasting with the horn poppy.

Glaucium flavum f. *fulvum* makes even bolder and more conspicuously bluish foliage than *G. flavum*. Where this plant grows wild I have not yet discovered. It was given me many years ago by the great gardener and collector, Cedric Morris. As drought becomes relentless, it has the energy to produce a long succession of translucent, burnt-orange, poppy-like flowers, followed, as our native is, by long, narrow curving pods reminiscent of hunting horns. These in themselves create a remarkable, almost cage-like construction from midsummer to autumn, which, I imagine, is designed to be bowled along by wind to disperse the seeds once it is detached by autumn gales.

AUTUMN BULBS

Most colchicums are over by mid-October, just a few stragglers remaining. True crocuses were late in arriving, waiting, possibly, for a little moisture to set off their time clock. Each autumn I anticipate the first flowers with the same excitement I feel for the first crocus in spring. Yet every autumn I am tormented with doubt, concerned the crocus may have been eaten by mice, but then I am relieved to see new ones arriving daily in increasing numbers, where single bulbs have made up into satisfying clusters, and small bulbs have fattened up to flowering size. I like to think this is the result of the pest control team – four resident cats doing their job and earning their breakfast.

A crocus I specially love is *C. speciosus* 'Albus'. It stands up so proudly, snowy white bowl-shaped flowers with sharply pointed petals, marching between snake-like sprawls of *Euphorbia myrsinites*, drifting between low tussocks of the American blue-eyed grass, *Sisyrinchium angustifolium*, through thyme mats edging the walkways – or just through bare gravel alone, where, on a bright sunny autumn morning with flowers opened wide in response to the warmth, I almost think they look best of all. *Crocus kotschyanus* var. *leucopharynx* is increasing well, its lavender flowers on thin white stems now making generous-looking bunches, in some places through mats of golden-leafed thyme, sometimes through bare stones, where I know *Nectaroscordum siculum* subsp. *bulgaricum* lies hidden deep below, waiting for spring.

Brightest of all the small bulbs are glistening yellow sternbergias. Although they look very like crocuses, sternbergias are more closely related to lilies (as are colchicums), since they belong to the family Amaryllidaceae, cousins of the

Liliaceae. Unlike crocuses, whose leaves do not appear until spring, sternbergia flowers appear clasped in clusters of dark green, shining strap-shaped leaves. The flowers and fresh-looking leaves together are most invigorating to come across at this dying time of year. They need a warm spot in well-drained soil to encourage flowers as well as leaves. Mine are growing in sun-baked soil along the edge of a west-facing border, along with the pretty parasol-shaped leaves of *Geranium cinereum* 'Purpureum' (dark purple form from Arends Nursery, Germany), wandering thickets of the dwarf Russian almond, *Prunus tenella*, together with a handsome feature plant, *Euphorbia rigida* – all these companions promising to enhance this space with flowers next spring.

Nerine bowdenii, such a lovely thing when well suited, continues to exasperate me, appearing only as isolated flower heads still. It is another member of the Amaryllidaceae family from South Africa, where it grows on screes and ledges in the Drakensburg Mountains. Driving past other people's gardens I see huge bunches of these exotic, sugar-pink, lily-like flowers, and turn quite green with envy. I suspect on enquiry I might find they had 'been there for years', ignored until they suddenly burst forth like a firework display in late October. Certainly they need time to establish, and a warm spot, protected from severe frost. In hard winters past, we sometimes lost bulbs completely so they would hardly be viable in heavy soils in really cold districts, especially since they like to thrust their bulbs halfway out of the soil, probably to receive a good baking, to encourage flower formation.

We have a small collection of other forms of nerine. None I think is reliably hardy. We grow them in clay pots and pans, overwintering them in a frost-free greenhouse. Emily Paston, who cares for them, brings them into the Gravel Garden just before they open. She groups them round the base of the Swedish cut-leafed birch, *Betula pendula* 'Laciniata', which stands alone in the gravel entrance to my house. Already the birch is turning colour, the gravel littered with curling yellow and brown leaves. Emily piles the pots on different levels supported by 'fallen' columns of empty pots and old ornamental bulb crocks into which drift a confetti of falling leaves, all making an unusual still life.

Nerine filifolia, as the name suggests, has almost hair-like foliage with spidery rose-pink flowers. *Nerine undulata* is similar, not quite so spidery, but with pretty crimped edges to the petals. We do grow this outside. Several hybrids in white, rose and mauvy pink attract attention, but the species *N. sarniensis* is outstanding. It rolls back bright scarlet petals glistening with fine gold dust when caught in bright sunlight, to show off clusters of long, scarlet stamens and pistil, thrust aggressively forward, like beaks. From my kitchen window I see visitors come to view this unexpected display and occasionally I have seen a pot

disappear towards our Sales Area from where Emily quickly reappears to restore one of her babies. A notice 'Not for Sale' would be sensible, but would not prevent disappointment. However, we do not have enough stock of these half-hardy bulbs, nor time nor sheltered space to propagate them commercially. Others do that. Generally I am reluctant to send plants to their death – plants which are not reliable outside in our climate. There are a few exceptions in the nursery but we try to make sure that customers know the risk.

I have already sung the praises of *Phyla nodiflora* – thyme-like in growth, smothering itself with bobbly heads of tiny flowers during late summer and autumn. There are a few flowers still. Surprising, even startling to some visitors, are clumps of snowdrops pushing their way through its carpet of green leaves. Some people feel discomfited seeing snowdrops flowering in autumn. They usually feel the same about autumn-flowering crocuses. It doesn't feel right, we're not ready for them is, I suppose, the feeling. But I welcome such *joie de vivre* when much else is bedding down for winter's rest. These fresh little bulbs lift my spirits and take my mind in a big leap forward to spring. This truly autumn-flowering snowdrop is *Galanthus reginae-olgae* subsp. *reginae-olgae* Winter-flowering Group (syn. *G. corcyrensis*)! It is found growing wild in Corfu and Sicily among rocks and oak scrub, where it flowers from October to December. It has increased well in my poor gravel soil, making good clusters of bulbs, with strong, healthy-looking flowers, unlike the form I originally had – *Galanthus reginae-olgae* (not the same), which always seemed a weakly creature with small flowers produced reluctantly in ones and twos. Perhaps I put it in an uncongenial spot, beneath a great oak, not imagining a snowdrop would be happy or look right out in full sun.

LATE AUTUMN HIGHLIGHTS

Overall the Gravel Garden looks well dressed in late November, creating scenes that entice me out in all but the worst winter weather. However wet and windy, it is a joy to walk here, not least because I need not put on wellies, since the dry gravel paths are always welcoming. Each way I turn I see integrated groups, companions, fitting comfortably together, as well-chosen, well-placed pieces of furniture bring life to an empty room. There is a wide diversity of shapes; some are large and bulky, like settees or wardrobes, others, smaller, might be compared with footstools or low tables. To lift the eye and paint the sky there are verticals, so essential in garden design, whether large or small. These include several different forms of eucalyptus, Mount Etna brooms and pencil-like junipers.

The highlights on late November mornings are the shadows, those invaluable dark accents which add so much to a design. The constant highlight, the focal point at this season, is the gravel, reflecting the light, itself warm and inviting me to walk into this fully furnished garden, albeit in muted shades.

The warm colour and glossy surfaces of bergenias make much needed contrast with the grey ghostly froth of *Artemisia ludoviciana* subsp. *ludoviciana* var. *incompta*. Much heavier shaded effect is made with the massed flat heads of *Sedum telephium* subsp. *maximum.* The brick-red flowers of *S.* (Herbstfreude Group) 'Herbstfreude' have weathered to the colour of dark coffee grounds while the white-flowered *S. spectabile* 'Iceberg' fades to a paler shade of brown, with bleached stems standing out ghostly white. Both these 'deads' could understandably be considered dreary by themselves, but in combination add something of value.

TOP *Bergenia crassifolia* 'Autumn Red'

BOTTOM *Bergenia* 'Mrs Crawford'

OPPOSITE *Bergenia* 'Mrs Crawford', the first bergenia to colour well, is more frost tender than some others.

A PASSION FOR BERGENIAS

Perhaps the dominant plant among all other foliages in the Gravel Garden is bergenia. I love the close-packed rosettes of cabbage-like leaves which spread into the gravel paths, pinning down the curve of a border or catching the eye in the distance, where repetition adds to the effect. Having survived months of drought, they create late autumn and winter scenes which I enjoy as much as the full glory of summer. We grow many different varieties showing different degrees of hardiness. Many of them, during the winter months, contribute various shades of mahogany, plum or coral-red, all polished bright as shoe leather.

There are gardeners who do not share my passion for bergenias, since the leaves can disintegrate and look very disreputable with the onset of cold and wet weather. More than twenty years ago I discovered Christopher Lloyd through greedily reading his classic *The Well-Tempered Garden.* Just one thing upset me. He had no time for bergenias. For the first time in my life I put pen to paper to write a letter to an author and take him to task for such a lapse in his otherwise overwhelming knowledge and experience in actually growing plants, not just writing about them. Back came a reply: 'Come to lunch.' So began a special friendship, based on many shared interests, particularly our passion for plants. But not always the same plants – that could become boring. 'Now Beth,' I can hear him saying, as we shivered round his garden at Great Dixter on a winter's morning. 'You can't pretend to love those wretched bergenias.' Poor things – most of their leaves had collapsed into sodden blackened heaps, leaving bare twisted stems covering little sprouting tips of new growth. I had to agree. They were a sorry sight.

Although I sometimes envy owners of good garden soil – that is, well-worked retentive loam, I know such soils, especially if watered by plentiful rainfall in winter, can be lethal to certain plants. Perhaps, in rich soils, bergenias would remain healthier in winter if planted on a raised bank where drainage could be improved. I sometimes drive past a churchyard where they were planted many years ago and now, allowed to grow undisturbed, great healthy clusters of leaves hang down the face of the retaining wall, attractive all winter, alive with pink blossom in spring.

There are numerous species available, having been found wild over a wide area, from Siberia (with what I do not know) or from sub-alpine meadows and edges of woodland in mountainous regions of Central Asia and the Himalayas. I imagine them on rocky slopes or mountain ledges, their roots tucked into some humus-filled crevice where they do not become waterlogged. I think my free-draining soil helps my plants (including hybrids) to retain good foliage in winter. I admit that a little more rainfall in summer would suit them better, that

PREVIOUS SPREAD The massed heads of *Bergenia* 'Schneekoenigin' (Snow Queen) almost smother clumps of overwintered leaves. If frost is forecast we cover the flowers overnight with fleece-covered hoops.

our conditions go almost beyond the limit of what they will endure. Yet on consideration I must contradict myself, since they have increased, thrived and given me pleasure for over thirty years.

I suppose it is a luxury (or madness) to be able to plant wide, sweeping patches of bergenia – some 5m/16½ft long by 1–1.5m/3½–5ft across – in the Gravel Garden. The old winter leaves will be buried in spring with blossoms, from white through shell-pink to deep rose and magenta, on rhubarb-red stalks, exciting contrast to the electrifying green of euphorbias. *Bergenia crassifolia* 'Autumn Red' is reliably hardy in my well-drained gravel soil but *B. cordifolia* 'Purpurea', with large wavy cabbage-shaped leaves, is possibly the hardiest and toughest, and generally a good all-rounder. Last year *Bergenia* 'Silberlicht' was the only form looking miserable after frost and snow left it very dishevelled, but the flowers soon made up for its temporary dowdiness.

Bergenia 'Mrs Crawford' I named after a friend who had collected it (so I understood) in the wild, in Kashmir, about ten years ago. I should have sent it to the RHS at Wisley for correct identification. There is still time to organize that! It is a distinctive plant, making attractive neat rosettes of leaves with finely picoteed edges. It is one of the first to be affected, in late autumn, by low night temperatures when individual leaves appear enamelled bright cherry-red, while throughout winter all the leaves become a striking purplish red. But this bergenia is less tough than most others. Severe frost does damage the leaves. I allowed Christopher, visiting the Gravel Garden in April, to point them out with glee, but the last laugh I think rests with the bergenia itself, for a week or so later the tatty remnants of old leaves were buried beneath compact heads of pure white flowers enhanced by pale green calyces and young fresh leaves emerging. Looking well behind the warm autumn-tinted foliage of *B.* 'Mrs Crawford', the fine *Helleborus* × *sternii* 'Boughton Beauty' draws admiring glances in spring when, not yet opened, large clusters of buds are stained almost the same colour as the bergenia, shades of maroon mixed with touches of lime-green. The strident, almost raspberry flowers of *Bergenia* 'Abendglocken' need the light apple-green heads of *Helleborus argutifolius* (syn. *H. corsicus)* to calm the scene (perfect complementary harmony, I suppose, for those who abide by the colour wheel).

Bergenia 'Morgenröte' (syn. Morning Red) produces in spring heads of open-faced rose-pink flowers on slender coral stems followed by an impressive second flowering in June, set off by a silvery grey carpet of *Stachys byzantina* 'Silver Carpet' nearby, vitally alive against the slowly darkening heads of *Sedum* (Herbstfreude Group) 'Herbstfreude'. *Bergenia* 'Rosi Klose', well named after the hardworking wife of that indefatigable plant collector, Heinz Klose of Kassel,

LEFT *Bergenia cordifolia* 'Purpurea'

RIGHT *Bergenia* 'Silberlicht'

north Germany, bears clustered heads of clear rose-pink open bells, attached to the main stalks by pinkish brown sepals and petioles – a lovely combination.

I grow other selected and named cultivars of bergenia, many of them recent introductions, as well as several species, including the small-leafed form *B. stracheyi* (with both pink and white flowers), and one with curiously hairy leaves called *B. ciliata*. To try to describe their individual contributions would be next to impossible for me to write and tedious to read. But anyone who has a passion for some particular species will understand the fascination I have for bergenias and the important role they play in my garden design.

TIDYING UP FOR WINTER

Surveying the scene with my girl gardeners who have come to 'tidy up' for winter, we stand back to analyze the overall effect, to judge what is good, and what is looking a mess. Plants that have disintegrated into a mass of broken sticks or collapsed into squashy heaps – like the lush leaves of *Galtonia candicans* or *Agapanthus campanulatus* – need no soul-searching to decide what should be done, but others need more thought, especially since I like to keep 'deads' that have good structure and often retain attractive seed heads. Together we make fairly serious decisions, such as to remove a plant or shrub which by now has fulfilled its original purpose and needs to go, to make room for its neighbours. Such is the case with a grey-leafed × *Halimiocistus* lying almost smothered by a ceanothus which has suddenly 'put on weight', having a spread now of about 2.5m/8ft, making a handsome dark evergreen mound in front of a group of silvery green variegated privet, *Ligustrum ovalifolium* 'Argenteum'.

When later I walk back to see how the girls are getting on I find the picture is more sharply defined – the fuzzy pieces have gone. It reminds me in a silly way of sitting in the hairdresser's watching shaggy heads being restored to neat and tidy shapes, emerging with a more youthful appearance. Gardens, too, can look elderly when unkempt; judicious pruning and grooming retain a fresh and vigorous appearance.

My two experienced gardeners Lesley Hill and Winnie Dearsley have been doing this tidying for over twenty years and are used to my puzzling instructions. I wonder what the two apprentices make of it all. It would be simpler if I could say remove everything which now looks sere or dead, but that would mean cutting down such things as *Phlomis tuberosa* 'Amazone' and the seed heads of several different kinds of origanum, or marjoram, some of which still look attractive with myriads of pale, shining stems and fuzzy brown seed heads. In contrast to the overwintering foliage plants, they create gauzy brown shadows, a complete change of texture.

The Russian sage has lost its leaves, but its ghostly white stems make lovely skeletons shaped by the wind, memorable when the sinking sun spotlights tiny gnat-like creatures dancing among them. These, and anything else which retains good shape, I say please leave for the effect they make now, and for later, when a frosty morning will transform every twig with ice crystals.

It is too soon to remove the cloud-shaped bundles of *Gypsophila* 'Rosenschleier' (syn. *G.* 'Rosy Veil'), as they still carry a late crop of tiny flowers. They are unlikely to be cut down till much later, since they too, without flowers or leaves, contribute a change of shape and texture against the gravel path. Each plant makes a wide circumference. At present, two plants are making a spread of about 2m/7ft. Eventually they will be reduced to small fist-sized crowns of resting buds. But by spring the space they leave will not be empty, but will be taken by spring bulbs.

Most of the alliums were tidied away long ago, but one of the latest to flower, *Allium carinatum* subsp. *pulchellum,* I find seeded in and around cushions and mats of helianthemum, *Veronica cinerea* and *Eriophyllum lanatum.* Self-sown, they created an almost meadow-like effect, delightful in flower, both pink and white forms, but their tumbled stems are distracting now. Once they have been cleared away, the area left, of gravel mulch, will look as fresh as a newly shaved chin.

The shabby-looking plants of *Genista lydia* that I was in two minds about earlier have been removed; so too a vastly spreading carpet of thyme looking tatty and threadbare after several hot summers in baking gravel. In spite of poor conditions this plant had taken over several metres of border edge, not only suffocating weaker neighbours but also providing cover for bulbs. We shall wait till spring now to see where the bulbs are, so that we do not spear them as we replant the area.

As I have said before, free-seeders can become weeds, that is plants in the wrong place. I have been grateful to my gardeners this week who have painstakingly taken out a rash of bupleurum seeded into a carpet of thyme. Most of the time the girls have to make such decisions themselves, to remove or to leave. When in doubt the rule is to leave. But sometimes as now, in late autumn, it is easier to see when something needs clearing out. Do it now, and make an empty space ready for when we are looking for a home for new additions.

WINTER PICTURES

Yet another day in late November with overcast skies, drizzling faintly, yet it is beautiful to stand here in the Gravel Garden, in the mild damp air, everything silent and still, in utmost calm except for the sound of birds, blackbirds and robins mostly, hidden in the wealth of foliage all around.

Curious as it may seem, the Gravel Garden is as furnished with colour, textures and shapes as it is the rest of the year; not with the fresh lime, white and yellow tones of leaf and flower in spring, not with the wealth of alliums, poppies and rock roses in high summer, nor with the highlights of late autumn flowers, but with good foliage and seed heads. It presents a tapestry of warm shades and contrasting textures – bright feathery grasses, lively green, silver and gold foliage, enclosed on the east side by the leylandii hedge, while on the west boundary the two huge and ancient oaks, russet now against the grey sky, give stature and a sense of timelessness.

At the end of December, past three o'clock, a silvery bar of weak sunlight peers through massed grey clouds illuminating the Gravel Garden. The Great Oaks throw long shadows across the gravel. Odd plants are caught in the splintered rays of sun. *Euphorbia myrsinites* sprawls like a huddle of blue scaled snakes, one writhing stem showing the first terminal rosette of pea-green flowers nudging the stones. *Euphorbia characias* subsp. *wulfenii* stands nearby, reminding me irresistibly of the feathered necks of big birds (must be the wildlife films) with ruffs of pewter-grey overlapping leaves lit with pale veins, glistening with raindrops. In some forms the topmost clusters are stained pale maroon. In others, the flower buds are already formed, 'eyes' downcast beneath haughtily bent necks.

Even in January with hardly a plant in flower, the Gravel Garden still presents a complete picture. The grasses hold good shape, caught in low shafts of sunlight; their pale shades of buff and straw catch the eye from far away, creating vertical shapes or softening the strong shapes and tones of evergreens such as hebes, euphorbias or cistus. Where dormant bulbs or herbaceous plants lie resting far below, the gravel mulch conceals the empty eyesore of dark bare earth, curving sinuously between tufts and tussocks of cut-down remains or making pale contrast against carpets of dark thymes or silvered lambs' ears (*Stachys byzantina*).

Most large sedums, cultivars of *Sedum spectabile*, still show up well, offering varying shades of brown domed seed heads making dark accents between silvery grey foliage. Outstanding is *Sedum* 'Matrona', taller and larger than all the rest, with heads the colour of dark coffee grounds. The seldom seen euphorbia, *E. rigida*, draws attention to itself even in mid-winter. Its conspicuously rose-tinted flower stems (30–38cm/12–15in tall) show up, since they have lost their lower leaves by now, and curve outwards and upwards to display terminal rosettes of pale pointed wedge-shaped leaves clasping in their hearts clustered flower buds as yet hidden from winter cold. New shoots for the coming season are already formed at the base.

DESICCATED SEED HEADS

It is my practice not to cut down herbaceous plants too early. Some must go because otherwise they would collapse into soggy heaps, but many that retain good shape continue to add interest, especially when rimed with hoar frost. It improves the effect sometimes to clear away raggedy leaves or crossing stems, but not too much. A well-preserved seed head can be just what is needed to frame a view.

It would be a sin to cut down the seed heads of the various cultivars of *Sedum spectabile.* They make necessary dark shadows above their succulent leaves, now turning melon-yellow before they fall.

The white woolly coating on the thick stems of *Verbascum bombyciferum* (standing like sentinels all summer inside the entrance to the Gravel Garden) was almost worn away by October. In winter their brown, hard seed cases close-set on long branching stems form carved candelabras. Sharply outlined against the sky or blown into twisted shapes, they create a framework for the view beyond. Beneath them are handsome rosettes of leaves, new plants gathering strength to make next year's display.

Tidied up a little in the autumn if need be, the stiff branching stems of *Verbena bonariensis* create valuable verticals until storm and tempest topple them. The light purple flower heads have turned dark chocolate-brown. Held on bare branching stems, they form a pattern of open screens. Like *Verbascum bombyciferum*, these plants self-sow among wide patches of low-growing plants, or lean unexpectedly over the gravel paths, their dark bobbly heads silhouetted against the pale stones. 'Wonderful plant that,' says a visitor just walking by. 'Makes such a show, takes up so little room.'

We remove the raggedy spent leaves of *Phlomis tuberosa* 'Amazone' after flowering so we can enjoy its small brown bobble-shaped seed cases spaced at intervals along rigidly upright branching stems, defying all the gales of winter. Through its 'wooden' screen, I enjoy the lively contrast of woolly-leafed ballota, the proud curved heads of *Euphorbia characias* subsp. *wulfenii* and sharp green mounds of *Santolina rosmarinifolia* (syn. *S. virens*).

LEFT *Euphorbia characias* subsp. *wulfenii*

MIDDLE *Euphorbia myrsinites*

RIGHT *Sedum* 'Matrona'

OPPOSITE TOP A few strong forms hold together a scene that is becoming blowsy, when most subjects are past their best. *Sedum* (Herbstfreude Group) 'Herbstfreude' pulls everything together with good shape and warm colour.

OPPOSITE BOTTOM LEFT *Bergenia* 'Abendglocken'

OPPOSITE BOTTOM RIGHT *Bergenia cordifolia* 'Purpurea'

SHAPES AND COLOURS

The overall appearance of the Gravel Garden changes little during the winter, appearing like a stage set-in-waiting, coming to life when pale winter sunlight illuminates a quiet scene of colour and form. It is always a welcoming environment to walk through, even on the dullest day.

I planted several different berberis which, rather to my surprise, took longer than I expected to establish themselves in this poor gravel soil with little rain and high summer temperatures. But now, after six or seven years, several bushes of *Berberis* × *ottawensis* f. *purpurea* 'Superba' have put on both height and bulk, while the normally plum-purple leaves have turned brilliant wine colour.

Different forms of bergenia exhibit varying shades of red, from dark plum to bright cherry and scarlet. *Euphorbia epithymoides* is bursting into flames of orange and red. Contrasting nearby are big patches of the lemon-leafed form of lambs' ears, *Stachys byzantina* 'Primrose Heron', and the pale green and yellow form of sage, *Salvia officinalis* 'Icterina'. The boldly variegated *Euonymus fortunei* 'Emerald 'n' Gold' looks striking beside the plain leaden-green leaves of *Cistus ladanifer*. (Have you crushed in your hand the slightly sticky leaves of this cistus and smelt its spicy scent, or wondered where that faintly evocative scent came from, drifting on a still, humid, summer-evening's air?)

Throughout this landscape of heights and hollows, grasses in flower, tall or short, gauzy or bulky, illuminate the scene in shades of buff, honey, orange and gold. Bush clover, *Lespedeza bicolor*, is not such a dramatically beautiful shrub as *L. thunbergii*, whose arching stems are bowed to the ground in autumn with elegant terminal panicles of small rose-purple pea flowers. The flowers of *L. bicolor* are later, paler and in smaller panicles, but overall the upright shape and texture of this shrub pleases me. In spring it is cut to the ground, then throughout summer produces many light cane-like stems topped with sprays of tiny leaves. Flower trusses of pinkish purple pea flowers are produced in the axils of the leaves. Frost turns the leaves brown, the bare stems olive-brown. From my kitchen window, lit by low winter sun or glistening with frost crystals, the dry columnar shape (1.25m/4ft tall) softened with wispy curling side branches swaying slightly in a light breeze, attracts my attention.

SHIMMERING GRASSES

On winter mornings, when frost is slowly peeled away by almost horizontal rays of sunlight, I delight in the effect made by grasses in my Gravel Garden, in their vital contribution to the scene. If I remove them in my mind's eye from among their neighbours it would be like turning out a light. In daytime and at night, when the car lights pick them out as I drive home past the Gravel Garden, I love the contrast of the grasses' gauzy shapes. To those who might be troubled to see dried grasses shimmering among live plants – polished bergenias, sword-like yuccas and felted greys – I can only say, just try some. It is not until you grow a selection of grasses that you begin to realize how individual they are, with such differing characteristics, different personalities. Yet when it comes to describe them I find myself stuck for words to make sufficient distinction between them.

The plant we call *Stipa splendens* (not yet finally identified) takes all summer to make a very impressive display by autumn. Initially it looked thin and sparse but, continuing to produce new flower heads from the centre, it now has made a beautiful wide fan of stems ranging from 1.25m/4ft to 2m/7ft, each topped with a delicately feathered head. To walk round a corner and come upon this fountain-like grass standing well above its neighbours and arching out over the gravel is for me a joy. In October, as I sit before it on my stool, it makes a gauzy screen through which I can see the Gravel Garden stretching away into the distance, jewelled with late flowers. Rooks flying overhead draw my eyes up and I can see, like a fine pencil drawing, the tracery of tall grass heads against a pattern of blue sky and thin cloud.

By the end of November the summer grasses *Stipa gigantea* and *Helictotrichon sempervirens* have done their bit. Their wind-damaged stems are cut down, leaving resting tussocks of foliage nestling unnoticeably between euphorbias, salvias and santolinas. Now there are other grasses which have developed over many weeks to be ready to play their part long after more ephemeral, two-pence-coloured flowering plants have left the stage.

Calamagrostis × *acutiflora* 'Karl Foerster' has become the colour of ash-blond hair, and as attractively straight as is the fashion of today. Each established clump forms a column of thin blanched stems topped with seed cases held rigidly upright. Illuminated by blindingly low sunlight, which leaves heavy dark shadows behind the grass, this picture looks as if it were lightly chalked on to dark paper, a pattern of light and shade. This strongly vertical form is perfectly partnered by another grass, gracefully feminine, curtseying alongside. This is *Stipa calamagrostis*, whose soft feathery flower heads create a fountain-like shape, ever increasing as new stems emerge from the centre, as old ones bow to the side. The flower heads are pale shining green when young, fading to soft fawn as they mature. All stages contribute to the overall effect of light and warmth as the massed stems sway above polished rosettes of bergenia, contrasting with the deep rose and brown heads of *Sedum* 'Matrona', all backed by the rich plum-coloured foliage of *Berberis* × *ottawensis* f. *purpurea* 'Superba'.

Holcus mollis 'Albovariegatus' is avoided by some gardeners because it can be troublesomely invasive in the wrong place, as among treasured alpines, but here I welcome the low tufted carpet, half-smothered in autumn with copper-coloured oak leaves. All summer it plays its part, a bright contrast in colour and texture to its neighbours. Creeping stringy rhizomes carry tufts of soft leaves with broad white margins centred by a narrow green strip, giving the effect of a white carpet with fresh foliage in spring, and again in autumn when much-needed rain will have encouraged new growth. Although invasive, this grass is easily dislodged and will succeed in shade or open places. I have planted one of my favourite snowdrops, *Galanthus* 'James Backhouse', in gaps left where this grass shrinks inconspicuously in winter, so repeating in February the green and white pattern which will be carried on when the snowdrops fade.

Poa labillardierei is well designed for dry soil. It makes a large neat tussock of tough grey-green leaves, which roll up like thin wire to help conserve moisture. Above them new flower heads are produced continuously throughout summer, forming a fan-shaped haze composed of countless stems, 1.25–1.75m/4–6ft tall. By autumn the delicate flower heads are bleached, filtering late afternoon sunlight and shimmering above drifts of *Colchicum* 'Rosy Dawn', which nestle among the deeply cut, parasol-shaped leaves of *Geranium macrorrhizum*, flushed in shades of crimson, orange and apricot. To give some idea of the stature of this grass I have planted it elsewhere in heavier soil, beside *Phormium tenax* Purpureum Group. Although the grass is half the size of the phormium, the two make an interesting complementary pair, each repeating the same fan-like shape, but opposite in texture, the phormium hard and sword-like, the grass its filmy counterpart.

The Japanese grass *Miscanthus sinensis* 'Yaku-jima' has amazed me by making a substantial feature in the poor dry soil of the Gravel Garden. It is now 1.5m/5ft tall and an amazing 2.75m/9ft across! It is not so tall as many forms of miscanthus, but, over about five years, vast multitudes of flowering stems have pushed up from the centre to increase its girth. They gradually fan out until the whole plant makes me think of a huge crinoline petticoat with layer upon layer of soft lacy brown flower heads interwoven with narrow ribbon-like foliage in shades of copper and tan. Lowly companions, wide-spreading mats of dark green thyme and whitened lambs' ears (*Stachys byzantina*) are happy to share in this glory.

Eragrostis curvula has caught my attention for months. Like many grasses it takes a while to build a strong tussock to provide the wealth of flower stems needed to create a good effect. A few spindly stems would be insignificant; an explosion of many hundreds is memorable. Although I could use almost the same words to describe its leaves and flower heads as I have for, say the poa or calamagrostis, this grass is distinctively different. Eragrostis is lower, and even more densely packed with hair-fine flower stems, each one illuminated against a background of shadows to create a scintillating sunburst effect.

The awkward name *Oryzopsis miliacea* I find hard to remember, but this grass makes a memorable feature in a corner where we enjoy it throughout summer, and then on into autumn and winter. Like many grasses, it looks best planted in a void, that is where its neighbours are all much lower so the exquisitely feathery heads on stems 36–48in/90–120cm tall have no interruption. Thymes, vincas and other low carpeting plants make good companions, while agapanthus (not too close), both in flower or seed, adds a touch of solidarity. This grass does seed around, especially in bare soil. We find it best to cut it down after the first flush, before it has seeded, when it will quickly refurbish itself to make a lovely feature well into winter.

The little blue stem grass from the short grass prairie of North America was *Andropogon scoparius* when I met it growing in the wild several years ago, but now botanists have decided it should be *Schizachyrium scoparium*. Christopher Lloyd and I were on our way home from giving talks in Melbourne and Toronto. We backtracked to Minnesota, where we were given the opportunity to see some familiar garden plants growing wild in the shamefully small, one-per-cent remaining virgin prairie – only small pockets left after millions of hectares have been swept aside for the production of corn and hogs. There are two types of prairie: the tall grass prairie, growing on very deep rich black soil supporting lush grasses and imposing perennials, including, I was fascinated to see, *Helianthus tuberosus*, which we grow in the vegetable garden as Jerusalem artichoke; and the short grass prairie, growing on thinner poorer soil. Seeing these plants in their natural setting, it was the grasses, caught in low sunlight, which enchanted us most, especially the little blue stem. It was everywhere, the prettiest thing to pick for dried arrangements, growing in billowy clumps, about knee high, with clouds of silvery seed heads held above needle-like stems, blue-green when fresh, but warm foxy-red in autumn. Other familiar grasses were the blue-grey June grass, *Koeleria macrantha*, a more emphatic plant for the border edge, *Festuca glauca* and *Bouteloua gracilis* (syn. *Chondrosum gracile*). Woven like a repeating theme into the coloured fabric of the prairie were smoke-grey bands of *Artemisia ludoviciana*. Shame on me! – there is as yet only one little blue stem grass in the Gravel Garden. Here, throughout summer, its very narrow blue-grey leaves form a narrow upright tuft 60cm/24in tall (it can be 70cm/28in in richer soil), looking well erupting from low mat-forming plants. It forms curious narrow heads of elongated seed capsules, while in autumn the whole plant assumes hazy shades of bronze-purple. In the wild, autumn colour ranges from bronze to flaming orange. I read that this grass can be a heavy seeder. While I prefer to recommend grasses that do not seed or run about invasively, I cannot condemn such habits entirely. We would deny ourselves the pleasure of many flowering plants if we did not take steps to prevent them becoming troublesome. And in some situations a meadow of little blue stem might be just what is needed.

Bouteloua gracilis is one of the main grasses of the short grass prairies. It has quaint charm grouped along the border edge, or used as a specimen among small plants. Stiff little flower heads, silvery white when fresh, developing into brown seed cases, are set at an angle on wiry stems 50cm/20in high, reminding me of the crumb brush from days when, as a child, I was asked to sweep the tablecloth. They are thought also to resemble mosquito larvae – hence the common name mosquito grass. The first frost turns the whole plant purple, which bleaches to straw colour by winter. It is tolerant of extreme cold and heat.

Blue sheep's fescue, *Festuca glauca,* is a remarkable accent plant on account of its powder-puff-like tussocks packed with narrow grey-blue rolled leaves. There are several named cultivars. *Festuca glauca* 'Elijah Blue' is probably the bluest form. *Festuca glauca* 'Golden Toupee' forms wide tuffets of luminous yellowy green tightly rolled leaves. The feathery flower heads stand 30cm/12in tall, making great impact among flat thyme mats. *Festuca valesiaca* 'Silbersee' is a small 10–30cm/4–12in compact form with silvery blue leaves. All these fescues need well-drained soil and full sun. Too dry conditions or too wet winters may cause dieback from the centre. Divide the plants and replant live divisions in spring, and trim the foliage of established plants to 10cm/4in above the crown to keep them tidy.

Helictotrichon sempervirens forms a beautiful sunburst shape, composed of arching, evergreen, vivid blue-grey foliage topped in midsummer with oat-like panicles of flowers, bluish white when fresh, bleaching to buff seed heads. The plants grow 30–45cm/12–18in tall and as wide, while the flower stems stand 30–60cm/12–24in above the leaves. Their eye-catching shape and colour remain a feature throughout the summer months. They are best planted singly, as accents to stand above lower companions. In groups the plants need space, to display the fine foliage and flower fully.

The large blue hair grass, *Koeleria glauca,* is not unlike festuca but has more substance. Its dome-shaped clumps of overlapping blue-grey leaves, up to 30cm/12in tall and across, are valued as accent plants, as an edging, grouped in the foreground of a mixed border or placed individually among alpines or small

Dried grasses shimmer among live plants in winter, creating a pleasing tableau of contrasting, gauzy shapes.

plants. It needs replacing about every two years with freshly divided pieces, in spring, to maintain the fresh effect.

The huge genus of fountain grass, *Pennisetum*, includes some of the most attractive flowers of all ornamental grasses. They mostly need hot summers plus sufficient moisture to produce a fine display in late warm summer–autumn. In my Gravel Garden I grow *P. orientale.* This grass needs a warm summer when dense hummocks of fine narrow leaves become crowned with fuzzy flower heads. They dangle, fat and hairy, like soft grey-mauve caterpillars suspended from thin wiry stems 45cm/18in high. *Pennisetum villosum* will not stand many degrees of frost, but is so beautiful it is worth the trouble to pot it and protect inside over winter. Above the grassy leaves there appears a constant display of furry white caterpillar-like flower heads, a silky greenish white when fresh in July–August, becoming creamy white as they mature. My plants weave between stems of *Gaura lindheimeri* and *Verbena bonariensis*, a combination that delights us and our visitors well into autumn.

One of the loveliest verticals, especially when caught in early morning or evening sunlight, is *Stipa gigantea,* which comes from Spain and mountain regions of Portugal. From dense, basal, evergreen tuffets of finely rolled leaves spring many tall stems, up to 1.75m/6ft tall, with loose panicles of oat-like flowers, metallic in texture. Long golden awns and golden anthers increase the shimmering effect, since all quiver in the lightest movement of air.

Mexican feather grass, *Stipa tenuissima,* has soft clumps of hair-fine foliage and an endless display of silvery green flower heads 60cm/24in high, each seed case tipped with a long silky filament. The overall effect of delicate plumes and fresh green foliage waving in the slightest breeze is magical, among daisy flowers especially. We cut off mature flowering stems in high summer to check seeding. These are not missed since fresh flowers are continuously produced until autumn, which look attractive until cut down in spring. Seedlings can be a problem, but are easily recognizable for removal.

SIGNS OF NEW LIFE

Already in the first week of January, there are signs of new life in the garden. We have had several weeks of mild weather with no feel of winter at all. The leaves of *Arum creticum* are well up, pushing through the carpets of oak leaves tangled among dry stems of *Euphorbia epithymoides* and the remains of *Phyla nodiflora,* dry and leafless now, waiting to be cleared away. Seedlings of love-in-a-mist and *Omphalodes linifolia* will rush in to fill the gap when the soil warms up. *Phyla nodiflora,* which I admire so much in late summer and autumn, will lie resting for some months before it chooses to give us a pleasant surprise later in the year.

The dense, grass-like clumps of *Ipheion uniflorum* create a bright green, almost meadow-like effect beside a swathe of bergenia, flushed now overall in deep wine tones. Proud stems of *Euphorbia characias* subsp. *wulfenii* are already uncurling clusters of pea-green buds. Too early, we cry! There is all of January and February weather still to come.

Galanthus 'James Backhouse' is emerging, opening its first flowers as soon as the buds of this sweetly scented snowdrop push through the stones. If it's too cold outside to smell them, you will catch the delicate scent if you pick them and stand them in a warm room. In the garden the stems slowly lengthen, till they stand 25–30cm/10–12in tall with long, narrow-sepalled flowers flaring wide open in winter sunlight, each flower almost 7cm/3in across from 'wing' tip to 'wing' tip, revealing the bright pea-green bow at the base of each rolled inner petal. Planted about four years ago, the bulbs have increased well. We must lift and divide them once the flowers start to fade, to spread this welcoming sight along the garden entrance border sheltered by the leylandii hedge.

TOP *Stachys byzantina* 'Primrose Heron'

BOTTOM *Galanthus* 'James Backhouse'

As I walk along the back of the border I look to make sure there are enough young plants of *Verbascum bombyciferum* to create grey-felted candelabras in appropriate gaps, to form much needed and long-lasting verticals to lift the eye above the rest of the planting. At the same time I find self-sown seedlings sometimes grown into sizeable plants, which must be removed because when full size they will cause damage to their neighbours. Some I leave, because I know I will enjoy the effect, perhaps on the edge of a path, among small plants, where I would not have thought of putting them and where the surprise element can sometimes be more exciting than the carefully considered effect.

In a mild winter it is important to watch for early shoots appearing in plants such as euphorbia and sedum, and to cut with secateurs, as low to the soil as you can, the hard spiky remains of last year's stems, which are both painful and damaging to encounter if you are obliged to fiddle among them, to make way for the new young growth.

Astelia chathamica (syn. *A. c.* 'Silver Spear'), with its beautiful silvered sword-like leaves, we protect with a roll of netlon staked around it, but such flimsy protection has been neither tested nor needed so far.

Needing no protection, the bush honeysuckle, *Lonicera × purpusii*, never fails to offer sweet scent from January to March, provided the weather is not arctic. In high summer it would scarcely be considered worthy of a prominent place in the garden, but in winter it deserves a place where you can regularly be cheered by it. Mine, needless to say, forms part of the structure of that favoured island bed alongside the driveway close to the house (see the plan on pages 28–9), where I can see its bare branches studded with fat green buds, each opening into pairs of creamy white chubby honeysuckle flowers, their clean sweet scent evocative of other spring flowers, of lily-of-the-valley perhaps or hyacinths. I am always tempted to pick a few branches for the house and do so, but the warm dry atmosphere inside causes them to drop quickly; they look better and last longer outside. When the flowers are over I like to prune the bush before too much new growth overtakes me, to prevent a dense twiggy effect.

NATURE'S FEEDING PROGRAMME

I am often asked how we fertilize plants in the Gravel Garden. Do we scrape aside the gravel mulch to add composted material, or do we sprinkle concentrated fertilizers? The answer is not quite straightforward. Most of the plants I have chosen for this area do not require extra feeding – would in fact grow out of character if over-nourished and therefore would be more likely to collapse when suddenly hit by high temperatures without adequate moisture to sustain them. Equally in winter, plants such as cistus and ballota will succumb to harsh weather, alternately wet and freezing, if they have made luxurious growth in rich soil, whereas those on a poor diet in well-drained soil will have made much more resistant, sinewy plants. Euphorbias, artemisias, most grey-leafed plants, bulbous plants and creeping thymes are found in stony places on hillsides in warm countries around the Mediterranean, or, like the humming-bird trumpet, *Zauschneria californica*, growing on dry banks along the roadside in California. These plants are not cosseted by man. Nature provides for them. They are adapted to a sparse diet. Because our 'soil' is initially so devoid of nourishment we aim, as I have said, to give fresh introductions a fair start, but after that they must fend for themselves. Most of them do.

The dying and decomposing of worn-out root systems, leaf fall and detritus blown in by the wind all become caught between the plants, trapped beneath the gravel mulch, pulled down and converted into plant food by a healthy population of earthworms, not to mention the millions (perhaps trillions) of bacteria inhabiting every cubic centimetre of soil. Of course, decomposing leaves can cause rotting, tucked among woolly foliage plants; we remove the excess, but overall they are left, together with any vegetable waste accumulating from the plants themselves, to break down gradually and feed the soil. Over-much tidying and clearing away of 'rubbish' is a mistake I think, leaving the soil hungry. In the past I often spread my bucket of weeds and cut-down remains beneath the skirts of shrubs or large plants, rather than waste time wheeling a laden barrow to some distant compost heap. (Others do it for me now, but I miss those days when I could work outside till dusk.)

So for the most part it is best to let Nature take charge of the feeding programme. But I do make a few exceptions. Where I have planted trees and shrubs – at the back of the 10m/30ft deep border alongside the leylandii hedge, and in the centre of some of the island beds – I still haven't the courage to add yet more gravel to an already stone-filled soil. Here, where it is less noticeable, we continue to mulch with straw. When the previous layer has disintegrated, we put down thick layers of straw in autumn after harvest when it is obtainable and has all winter to bed down. From time to time, we add well-rotted farmyard manure to assist shrubs I am particularly anxious to encourage. These include amelanchier, abutilon, specimen conifers, robust perennials like crambe and cardoon, and especially foxtail lilies, *Eremurus*. In my garden foxtail lilies need enough moisture and nutrients in spring to feed the long strap-shaped leaves, which build up the fleshy, starfish-shaped, brittle tuberous roots.

PRUNING AND SHAPING

We have been hoping for a snowfall, or good hard frost, to complete our year's collection of photographs, but so far neither frost nor snow has lasted for more than a few hours. We have been careful to cut down and remove only broken pieces or untidy crossing lines, which can ruin a scene when caught on camera, and to preserve everything which retains good structure to be ready for good hoar frost or snow-dusted pictures. Hoar frost occurs when fairly mild damp weather is suddenly checked by a sudden drop to freezing temperatures. Then the humid air is captured to form a thick coating of ice crystals, outlining every leaf, twig and blade. It reveals designs where there appeared to be nothing but a confused muddle before, and makes you observe the structure even of dead remains, whether angular, curvaceous or linear.

The mild January weather has enabled other necessary pruning to be done in comparative comfort. The shrubby mallows, including *Lavatera × clementii* 'Barnsley', which make vast growth in one season, are cut back now by about half, to prevent wind rock. They can be damaged by frost, but usually the woody framework is unharmed. They will be reduced further in March, to 75–125cm/30–48in from the ground, after which the strong root system formed last season will quickly burst forth new growth to form a well-shaped bush, over 2m/6½ft tall by midsummer. In some instances it is better to start again with a young plant, as they are so quick to fill up their allowed space.

Among the earliest shrubs needing attention are the brooms *Cytisus* and *Genista*. You can leave them as they are and they will carry on regardless, but brooms can become unwieldy, or simply too large for the space. As soon as possible after flowering is the best time to prune them. *Genista lydia* when young makes a nice tidy dome, clothed to the ground with fresh green shoots and leaves so tiny and narrow they are hardly noticeable. This is to reduce transpiration, an adaption to dry conditions. If not pruned regularly, including cutting out worn-out pieces, you will find yourself, as I am now, looking at a base of twisted, leafless, woody stems, with a covering of green shoots over the top half. It looks ugly. There is only one thing to do: have it out and replace with a young plant. Or use the space for something else.

Each year I like to look round the young tree brooms, *Genista aetnensis*. They need careful shaping (because they will eventually make a fair-sized tree), to ensure that they make a strong framework, with not too many main stems, which may split and spoil the shape. We thin out surplus branches, to ensure a single trunk to support the fountain of arching branches, which will be laden with flowers in July. We also remove low drooping boughs, which may whip plants growing beneath them.

The unwanted suckering shoots around *Rosa spinosissima* 'Falkland' have been removed and potted up, ready for the visitors who will ask for this old-fashioned-looking rose when it is in bloom.

From time to time 'old men' need replacing. Among these are woody, worn-out specimens of *Salvia officinalis*, various forms of santolina, helichrysum and some euphorbias. Because the soil is so free-draining, we are able, almost any day of the winter, to go out and remove the old fellows, take out some of the stony soil beneath them into a barrow brought alongside, and replace it with a top spit of improved soil incorporating well-made compost. These areas will then be ready to take healthy young plants in spring, including more bulbs such as lilies or alliums, which we will have growing in pots ready for planting. This spot treatment does not mean I can introduce plants that need a much richer, moisture-retentive soil, such as delphinium, astrantia or rudbeckia. None of these plants would survive in the Gravel Garden, whatever we did to improve the soil, when the temperature hangs around 26°C/80°F, and we often have eight weeks in high summer without measurable rain.

THE SCREE
GARDEN

Among the few problems that have arisen in the Gravel Garden over the past seven years has been the difficulty of accommodating small choice plants requiring a sunny site and well-drained soil. It was largely a question of scale and maintenance. As the garden matured, we were glad for some of the more boisterous shrubs and perennials to do their share in caring for the 0.35 hectares/¾ acre, but many small sun-loving plants had become overgrown.

Changes in the nursery and plans to build a small teahouse near the entrance to the Gravel Garden presented us with an unexpected site for a new garden that could provide a fresh home for small treasures. The Mediterranean Garden we had made there almost forty years ago was in need of an overhaul. Gardens, like their owners, become elderly. New ideas, involving moving plants around, inject fresh interest and delight.

Although I have been gardening for fifty-six years (the length of time we have been married), I can still feel daunted by the sight of a large empty space waiting to be transformed. There are always so many possibilities and permutations and in this case I wanted to make something quite different, using predominantly small plants and some of the easily grown alpines we have collected over many years. Visiting the Chelsea Flower Show more than forty years ago, I learnt to recognize and appreciate the skill and knowledge of small specialist nurseries from Scotland and other parts of the British Isles where alpines are the obvious thing to grow, who brought little tabletop stands to the show, displaying their treasures like gems on a jeweller's tray. I have little experience of growing true alpines, many of which need similar conditions to the high rocky places for which they are adapted, and had learnt when walking in the High Alps that conditions can be deceptive – plants apparently baking on rocky slopes are often being fed by underwater snow melt from the glaciers above. We could not emulate that. Fortunately for all gardeners there is a wealth of beautiful and interesting small hardy plants which are easy to grow. Many will tolerate, or even prefer, well-drained soil, while others are specially adapted to survive drought conditions.

To make a rock garden, however, here in flat Essex farmland (using imported Yorkshire limestone!) was out of the question. I have scrambled across scree slopes in the high mountains of Corsica and Switzerland, where shifting piles of rock fragments, shattered by frost, often dry at the surface, provide deep root-run for plants that have survived being swept away by avalanches. My gravel soil is not dissimilar in texture even though the terrain is flat. The five beds in my Scree Garden are not an attempt to create an alpine garden but are to show how to create maximum interest in a small area over a long period.

GROUNDWORK

The area we chose for the Scree Garden, approximately 30m/100ft long by 17m/60ft wide, is on the south side of our low, split-level house, where a wall runs north–south, parallel to the house, dividing the garden from the nursery. Within this wall our original Mediterranean Garden surrounded the south-west part of the house. I decided we could take down a section of the wall – just enough to tuck in our wood-framed greenhouse where it would be both convenient for the nursery and attractive as a background for the new garden.

Midsummer was a good time to begin dismantling the greenhouse, since very few plants require shelter there during the hot months. David Plummer, our able handyman, carefully removed all the glass panes, strengthened the woodwork where needed and finally unbolted the wooden framework from the block walls on which it rested. Meanwhile cement block walls were erected on the new site. On Thursday 2 September 1997, at 8.30 a.m. I made sure of being outside in time for the big event, to see eight of my men, including summer students, transport the framework of our 35-year-old greenhouse to its new home. I watched them lift the structure, supported on four strong timbers, moving backwards and forwards to find the right angle to set it down on its new base. The operation went perfectly, all men straining with the weight but making it without a hitch. I was proud of them, and delighted with the result: the greenhouse was situated much more effectively than before.

We stripped out everything worth salvaging from the old garden. Already David Ward had rescued rare fritillaries as they died down after flowering, together with other small treasures from the original small gravel bed; he potted them on and tucked them away in part of a plastic tunnel we call 'Pets' Corner', an area reserved for newly introduced plants waiting for a home. Cuttings were taken where needed, but anything already growing in the big Gravel Garden was not duplicated. The next task was to remove the top spit of soil because over the years it had become infested with 'weeds'. Some, like *Crocus tommasinianus* which seeds everywhere, were plants in the wrong place. Others, like celandines, had been brought in from elsewhere, where a few wretched little bulbils had found their way into the compost heap. The soil removal was done with a scoop bucket on front of the tractor.

The only feature left from the original planting was a Judas tree, *Cercis siliquastrum*, standing about 5m/16½ft tall in the centre of a flat area, roughly the size of a tennis court. During the winter of 1997 I watched this space, empty except for the Judas tree, with mixed feelings, missing the security of what had been familiar, knowing that time and effort would be needed to create something comforting, close to my house.

DESIGNING THE BEDS

I knew it would take time, both for my design ideas to materialize and for my staff to be spared from the vital work of the nursery to undertake the hard and heavy landscaping work. One morning in midsummer 1998 I suddenly was motivated to make a start. I assembled all our long yellow hosepipes and pulled them around, trying to clarify the design still germinating in my mind. The old garden had been based on straight lines, straight paths, rectangular beds. I wanted to eliminate this pattern and to make gentle curves to echo the shapes of the mature planting already there, but initially it was not easy to combine this idea with the straight lines of the greenhouse and recessed wall. By lunchtime a pattern of rounded beds with paths between them was evolving, but I was struggling with the Judas tree, which had been left on a small island of soil so as not to damage too many roots. David came out to see how I was getting on, and taking the hosepipe from me gave a few flicks, creating a much larger bed than I had envisaged, and suddenly the whole design fell into place: simply five irregular islands, easy on the eye, inviting to walk round. What a relief!

By July 1998, the fever of nursery propagation having abated a little, picks, mattocks and stout forks were brought out to dig deeply the hard stony subsoil of these island beds. After checking in a textbook on how to create scree beds it was obvious we already had plenty of good drainage – nothing but yellow sand and stones. Our need was to add enough improved topsoil in place of the soil we had removed. Once more the tractor was useful for bringing in the materials – long journeys with wheelbarrows back and forth across the nursery are a waste of time. We had a mixture of home-made compost, bonfire waste and finally clean topsoil, removed from the site of the new teahouse. This contained plenty of grit and stone to ensure good drainage throughout the new top layer we were creating. Initially the new beds appeared much higher than the surrounding paths, making me think of dogs' graves, absurdly out of place on this flat piece of land. (Was I making a big mistake?) I had originally toyed with the idea of making raised beds to show off small alpine-type plants but could not accept the idea of so much intrusive brickwork making a series of compartments looking like square boxes or round well heads!

There is a slight fall to the west, towards the rest of the garden, which slopes down to the water gardens lying in the hollow below. Gradually I could see an improvement if we tilted the beds slightly, shifting the soil about until we had made a series of gentle slopes. To support them I had in mind curbing which would vary in height according to the level of the bed, thus forming 'eyebrows'. I explained this with a rough sketch, the curbing higher at one end, tapering down to path level. We rang our local town council to see if they had a depot for broken paving and asked if we could buy some. They had and we could, so we collected it by the vanload, quite reasonably priced. Naturally if you have good local stone you would use that, but sand and gravel are our local stone, and using concrete made from these materials seems to me not an inappropriate thing to use here.

The next step was to bring out the hosepipes once more, to check every curve in relation to its neighbour, and to check the lines of the paths overall, some being narrower or wider than others, to make an easy flowing design.

HARD LANDSCAPING

The first wall to be built retained the long gently curved border alongside the greenhouse and nursery-boundary wall. It was the most important line to get right initially, since it had to relate to the free-standing beds alongside it. First a shallow trench was dug and filled with a cement mix to form a base. Next slabs of paving, cut to fit, were laid, four layers high. This height seemed about right in relation to the base of the greenhouse, to retain a narrow bed just high enough to allow some plants the chance to flop. Care was taken to brush out cement showing at the front of the wall, since I hoped the construction might resemble a dry stone wall when it was finally finished and weathered.

The island beds were contained in this way, except the walling was not the same height all the way round but consisted of perhaps three or four layers of broken paving to support the top of the sloping bed, which reduced gradually to one layer at the lowest end where ultimately the curbing would almost merge into the gravel path. Great care was taken, using a spirit level, to see that all the layers of walling were level. David cut the final top layer of pieces to shape, roughly rectangular, to avoid having sharp points and angles, so as not to look like crazy paving. Possibly we fussed too much – it may all well be hidden after a year's growth – but bad design or workmanship cannot be disguised. Hard landscaping must look attractive by itself to give real satisfaction. The view from the house of simple curving shapes provided just that in late autumn 1998, with nothing yet planted.

Without having paid attention to the paths, I still found myself contemplating five large 'pie dishes' standing proud. I could not rest until the paths were brought up to the right level. Having removed more than a spit of soil overall we needed a lot of material. First we brought in dumperloads of our own stony soil removed from digging trenches elsewhere. Then we ordered hoggin from our local gravel pits. Hoggin in our district is a bright orange, sticky combination of yellow clay and sand used as a base when making gravel paths. This was raked all over the pathways, and then well trodden, as you tread a bed ready for sowing onions.

Next 12mm/½in washed gravel from the same pit was barrowed in, Keith Page teaching how to spread by tipping and turning the barrow so the stones fall out in graded heaps rather than one big heap which is heavier to rake flat. Finally the gravel was rolled with a heavy hard roller, to bind the stones into the hoggin. When all this was completed I could breathe a sigh of relief. The dogs' graves and pie dishes had vanished. The beds tilted gently westwards and the raised path level reduced the height of the walling so that spaces between the beds had become pathways rather than passages.

Now I had to find time in the coming winter months to sort out lists of plants suitable for this site and to make sure we rescued the plants from Pets' Corner, which had been waiting too long to put their roots down into fresh soil.

WAITING TO PLANT

On a beautiful autumn day, in mid-October 1998, after night frost, the sky clear and blue, the air cold and still, the sun warm on my back, I sat in my empty Scree Garden, enjoying a blessed Sunday silence, apart from robins twittering all around and a fussy moorhen in the distance sending out her alarm call. I was content at last with the curved outlines of the new beds, with the way the eye is gently led around the stonework defining the edges, which contrast in tone and texture to the pale shining gravel of the pathways.

The Judas tree dominating the centre, laden with dangling seed pods, cast a network of shadows across the clean gravel. 'Here I am,' I wrote that day, 'at the beginning of a new project, the groundwork accomplished to my satisfaction, the winter before me, in which to decide how I will plant in the spring.'

For much of the country the winter was mild and wet, far too wet for planting. But *we* never seem to have excessive rainfall, whatever may be happening elsewhere. January was our wettest month, with 678mm/2.67in. February and March were messy with 185mm/0.73in and 333mm/1.31in respectively. We had very few frosts. I could have begun planting in March, but still I procrastinated. Part of my excuse for not tackling this new canvas was being preoccupied with the preparation of our new catalogue.

By the beginning of April 1999 I could make no more excuses. There had been no hard weather to retard growth and the season was already two or three weeks in advance. Every pot in the nursery showed a haze of new growth on top, while I could see tiny spears of white roots hungrily searching for a new home when I tipped out each plant. Just the right time to plant – to be able to see what you are doing, and where you have planted. I am never happy planting

All the hard work of preparing the site was done by hand, with picks, forks, shovels and sweat, to break up the compacted gravel, remove some of it and replace with a decent depth (about two spits) of improved soil. (Do you remember the barrowload of 'soil' on page 13?) Broken paving slabs were recut and carefully laid to make an edging of varying heights. The beds were tilted slightly towards the west. Starting to plant was both exciting and intimidating. There were endless permutations in putting plants together. This first bed has become home to Emily's collection of sempervivum, together with choice fritillaries and creeping plants.

balls of bare roots, when I cannot visualize what will happen next. When it came to the actual planting, I felt the need to make slight variations in level in the sloping beds, to create pockets for special specimen plants.

Tucked away in a corner, we had some pieces of rock-like substance, locally called ironstone, or ragstone. This is formed in gravelly soils when a pocket of clay forms a saucer, and the oxidation of iron in the water cements the gravel and sand lying above the clay into a rock-like layer. On farms it is broken up by a powerful subsoiler and brought to the surface. Left lying in the open, these irregular pieces of 'rock', excavated from my husband's fruit farm many years ago, had weathered, providing a home for lichens and mosses. All winter I had been in two minds about introducing the ragstone, so different in colour and texture from my broken pavement edgings, although both satisfied my principle of using materials local to this area. I convinced myself the edgings were already weathering. The ragstone would not be placed like almonds on a Dundee cake and would become acceptable once both materials were weathered.

The only problem was that we did not have enough ragstone. But we have a helpful farming neighbour. After a brief phone call we were away to pick up more heavy pieces of ragstone, overgrown with brambles, hidden alongside the farm's headland. We chose the largest pieces; too small pieces can look like currants in a bun. One specially fine piece needed three men to lift it. Even when properly placed and sunk it has an impressive 'cliff-face'. I call it Table Mountain – such is my foolish pleasure, since it may eventually be lost in the plants which surround it.

PUTTING PLANTS TOGETHER

Beginning to plant is always daunting, faced with so much empty space, which rapidly becomes too little for the wide choice of plants we have been propagating and developing for the scree beds. The only way I can plan is outside, on the site, with sheets of cardboard covered in lists, which I may have been making for several years, when whatever task in hand has put ideas into my head. Working on the catalogue always reminds me of plants not yet seen in the garden, as does going through our overloaded bookshelves, determined to dispose of some of the books – inevitably I end up keeping most, since I always discover something I had forgotten or never knew I had.

During the winter I had classified the plants according to size and habit, including verticals, accent plants, creepers and sprawlers, and, finally, bulbous plants. (These are listed on pages 184–5.) Now I began by picking up enough accent or dominant plants for all five island beds, primarily on the central raised areas. I had made a selection of shapes that complemented one another and would provide structure all year round. Before placing them on the beds, I submerged every pot in a bucket of water until the bubbles of air ceased and the pots felt heavy. Between these accent plants I placed flat creepers such as raoulias and thymes, which would cover the soil until the main plants had taken up their space. Strong verticals like *Libertia ixioides*, or the fine blue grass *Helictotrichon sempervirens*, I repeated, walking round the beds to assess how the eye would pick them up, often outlined in a void, or seen against the gravel path. Next came the small bushes and compact plants, including various small sedums, which would stay where I put them, without upsetting their neighbours, providing both flowers and interesting foliage. Finally, between and all around came the low sprawling plants, which I anticipated would spill over the low rim of the edging blocks.

During the winter our talented young student Yuko Tanabe, who was working for her degree in landscape design at Writtle College, had come over with her tutor and measured the area, preparatory to making an accurate drawing of the whole Scree Garden. Yuko, who often comes to help us, planned to enter all the plants. This would be helpful in many ways, not only to give a record of the original planting, but also to show changes over the next year or two, since change inevitably there will be. I never get it entirely 'right' the first time. Some plant associations may not work. Ebullient plants must be watched, and doubtless some removed before they cause trouble. Some plants may be too invasive, even though I deliberately include scramblers and sprawlers to give a softening, mature effect. Many of those conform to a little discipline, pruning or trimming at the right time, but usually the plants themselves sort out the stayers and the leavers. Where I plant thymes or carpeting phlox to make almost immediate ground cover between shrubby plants, I expect to see them smothered as the main characters take up their space, whilst elsewhere I hope to enjoy the sprawlers softening the low paving blocks by spilling into the gravel paths.

The initial planting completed, it is a delight to see these five beds already filled with colour and foliage, with promise of more to come. In the early morning, it is a joy to walk round, to see what new has happened, as exciting as peeping into the cot of the new baby. But these walks are not merely sentimental, other forces are at work – mildew, aphids, thrips, leaf mining insects – all must be noticed and dealt with, and weeds must be removed, to ensure the young plants will grow healthily in their new home.

SMALL SEDUMS

The small sedums, or stonecrops, are among my best-loved drought-resisters when the flowers have faded, a mosaic of contrasting leaf shapes, textures and colours is one of the most pleasurable effects in the garden. Almost all have evolved in mountainous areas where they flourish in the most meagre of soils, some even managing to survive on rocks and hence the name sedum, from *sedere*, to sit. The following I have planted in the new scree beds.

Sedum acre 'Aureum' is better behaved than *S. acre*, which grows wild throughout Britain. As a child I delighted to see cushions of this bright yellow-flowered wall pepper sitting on roof tiles or the tops of crumbly walls, and nowadays admire it growing along the dusty verges of dual carriageways, its seed scattered by the draught of passing traffic. It can become a rampant weed, smothering small edging plants, but *S. a.* 'Aureum' is well worth having when in spring its soft creamy yellow succulent leaves resemble (from a distance) a lightly cooked pancake tossed over dry gravelly soil.

Sedum aizoon 'Euphorbioides' should really be included among the herbaceous sedums (see page 117), but I use it to add a change of height among the scree plants. Standing 30–40cm/12–16in tall, it carries heads of warm yellow starry flowers that fade to handsome bronze seed heads on reddish stems contrasting with green toothed leaves.

Sedum hispanicum, from the Mediterranean, is similar in habit to *S. acre*. It creeps between stones, spreading mats of powdery-blue leaves the size and shape of rice grains. In times of severe drought they flush to shades of coral, encrusted with tiny pink and white starry flowers in June, like icing on a cake. *Sedum hispanicum* var. *minus* 'Aureum' is also similar to *S. acre*, but comes from south-east Europe. It is a neat-growing little plant, with tiny, soft creamy yellow succulent leaves, standing out between low grey or bronze foliage plants, and is attractive between sempervivum.

Here I include *Rhodiola heterodonta* (syn. *Sedum heterodontum*) and *Rhodiola rosea* (syn. *Sedum rosea)* since they are akin, and associate so well with small sedums. Both make a fat fleshy rootstock, dormant in winter. In spring tight pinkish bronze buds slowly produce a bouquet of radiating stems, 25–35cm/10–14in tall. *Rhodiola rosea* has blue-grey leaves topped with sharp lime-green starry flowers bursting with yellow fluffy anthers. In *R. heterodonta* the waxy leaves are almost iridescent with shades of lilac, amethyst and blue, on grey. Bronze-pink flower buds open to form a tuft of rich rusty-red stamens.

Sedum oreganum improves the drier the conditions become. Its close-packed clusters of fat fleshy leaves turn rich bronze-red in summer and its flowers of small, bright yellow stars on 15cm/6in stems fade to a quaint pink.

To break up the plains of these carpeters I might use the American blue-eyed grass, *Sisyrinchium angustifolium.* Two selected forms – S. 'Californian Skies' and *S. idahoense* – are both impressive, their tuffets of short grassy leaves topped with satiny, deep blue flowers, produced for weeks on end in midsummer.

Sedum spathulifolium is found wild in the Cascade mountains and coast ranges of western North America. It forms dense carpets of fleshy, purple-red rosettes, powdered with a waxy bloom. *Sedum spathulifolium* 'Cape Blanco' is identical in form, but the fleshy rosettes are silvery grey in winter becoming blue-grey in summer, overlaid with a white bloom, which intensifies with drought. Both these forms have soft yellow flowers.

Sedum spurium, a Caucasian species, is an invaluable weed-suppressing carpeter. It forms tangled mats of string-like rooting stems clothed in thick, oval, wavy-edged leaves. It will tolerate very poor conditions, even shade. Various selected forms include *S. s.* 'Atropurpureum', which has extra-good wine-red foliage, the colour intensified in high summer when smothered with clustered heads of shining starry flowers, intensely rose-red. *Sedum spurium* 'Green Mantle' I value for the fresh effect it makes with bright green rosettes of fleshy leaves forming dense ground cover on poor soil in open situations. *Sedum spurium* 'Tricolor' has trailing stems studded with rosettes of succulent leaves variegated with pink, cream and green, topped with clusters of pearly pink flowers in July.

Sedum 'Vera Jameson' is a delightful low-growing sprawly plant named after the garden owner where it was found as a chance seedling. Short, gracefully arching stems up to 30cm/12in long carry bloom-coated fleshy, purple leaves, topped with heads of dusty pink flowers in late summer. A most fortunate hybrid, introduced by Joe Elliott, of Stow-on-the-Wold, *Sedum* 'Bertram Anderson' is a variant of *S.* 'Vera Jameson', one could be substituted for the other. The difference is not very apparent. These low lax plants have dark stems, dark plum-purple leaves and terminal rosettes of starry wine-red flowers. Good edge-of-the-border plants, flowering late summer into October, the whole dark effect is very good beside a pale plant such as *Gypsophila repens* 'Rosa Schönheit' (syn. Pink Beauty).

There are heaps more sedums, hopefully to be found in specialist nurseries. I read about them in Will Ingwersen's *Manual of Alpine Plants.* All sedums are easily grown in full sun, in most soils except those with poor drainage or in excessively wet climates. They are invaluable in areas of low rainfall being naturally adapted to dry conditions. However, they are vulnerable to attack by vine weevils, little black beetles whose horrible cream, brown-headed maggots feed off the roots underground. No safe or reliable insecticide is as yet freely available to tackle them. In the garden we dig up affected plants, wash the remaining roots well and replant them in fresh soil.

A TRANSFORMATION

It is Sunday morning, at the beginning of July 1999, and I sit once more in my new Scree Garden as I did last autumn, when I surveyed the empty beds with only half-formed ideas drifting through my mind, trying to imagine how they might look when planted. Now I am relieved and astonished to see the transformation. The beds are overflowing with different shapes and textures of foliage and alive with colour. It is hard to believe there were no plants here only three months ago. Admittedly I have an advantage, being backed by the nursery, which enables me to use a broad palette of plants, whereas in years gone by (like most gardeners) I made do with less. But even in those days I increased my stock by seed or cuttings – and that was an experience that led to the beginning of the nursery. The way this latest planting has burst into life convinces us our nursery plants are truly viable and should go ahead well, provided the soil is adequately prepared to receive them. Indeed, as David reminded me, we have proved that less promising plants will take off under favourable conditions, since several of our new introductions had become decidedly sub-standard, restrained too long in pots. But with their roots teased out, it has been a joy to see them recover and flourish.

Unlike the dry, baking summer of 1995, the past three months have been comparatively cool and damp; with an average of 24mm/1in of rain per month from February to May, when we had 19mm/¾in, plenty of wind and enough warm days to make us pay close attention to the weather forecasts. To our relief, thunderstorms brought 64mm/2½in of rain during the first two weeks of June and a further 24mm/1in fell during the last week of Wimbledon. Together with the improved soil in the scree beds, that amount of rain has resulted, I have to admit, in uncharacteristically rapid growth – hence the fully furnished effect, as if the plants had been there for years! But this will change. The humus content will diminish and the plants will adapt to leaner times as they occur. Grooming and pruning will control some of the exuberance of this first favourable season. Some plants may well be removed to allow space for those less vigorous, or because I consider a colour combination could be improved. Already I have whisked away the lovely *Diascia barberae* 'Blackthorn Apricot' from a predominantly pink, blue and purple combination, and found a niche for it among yellows and silvery grey. Meanwhile I revel in a cornucopia of foliage and flower, after the previous summer when the site reminded me of a builder's yard.

Creeping plants like the blue *Convolvulus sabatius*, *Gypsophila repens* 'Dubia', *Artemisia glacialis* and various forms of mat-forming phlox are already sprawling over the low retaining walls, blurring the edges. *Raoulia tenuicaulis*,

TOP *Leontopodium wilsonii*

MIDDLE *Helichrysum* 'Schwefellicht'

BOTTOM *Eryngium variifolium*

from New Zealand, has crept over the summit of a bed chiefly occupied by a collection of sempervivum, small sedums and edelweiss, reminding me of the wind-cropped pastures of very high uplands. If we do not restrain the raoulia, more than likely an extra sharp frost, combined with wet, will damage some of it, but it has survived with us on stony soil, for many years.

Interesting verticals draw attention to themselves, such as *Nepeta tuberosa* with its upright woolly 'candles' formed by whorls of felted purple flowers and darker calyces. Rose-pink *Salvia* × *sylvestris* 'Rose Queen', a dwarf form of angel's fishing rod (*Dierama* 'Puck') and an unending succession of white stars on *Asphodelus fistulosus* all lift the eye above low mounds of thrift, various erodiums, hypericums and other compact plants.

Diascias make a blaze of rose-pink, both *Diascia rigescens* and *D. vigilis,* while *D. barberae* 'Blackthorn Apricot' is an orange-pink – salmon, perhaps. It now looks well beside the grey-leafed *Sideritis syriaca* with its spires of pea-green calyces hiding tiny lemon-yellow flowers, or with the distinctive *Sedum aizoon* 'Euphorbioides', a plant I much admire.

More soft yellows come with the glistening papery buds of *Helichrysum* 'Schwefellicht' (Sulphur Light) opening into tiny yellow daisy flowers. Just at this stage, mostly in bud, they dry well, hung up in an airy place. Standing out among many ashen-foliage plants, *Santolina rosmarinifolia* subsp. *rosmarinifolia* 'Primrose Gem' carries pale lemon button-like flowers above fresh green, finely divided foliage.

Pinks, blues and purples abound. *Penstemon* 'Evelyn' was for many years a resident in the old Mediterranean Garden on this site, but had become weakened and overlaid by coarse neighbours. Now, in refreshed soil, she is full of vigour, her slender rose-pink spires of narrow tubular flowers standing upright behind a bouquet of pale pink dandelion-like flowers of *Crepis incana*, while *Geranium* × *riversleaianum* 'Russell Prichard' with vivid magenta-pink, silk-textured petals sprawls into the gravel path below them. (David and I have already decided this lovely plant is taking up too much space.)

Grey- and silver-foliage plants already add to a sun-baked effect, and will provide contrast of texture and form throughout the year. They include the almost white-foliaged *Helichrysum stoechas* 'White Barn', a dwarf ballota collected in Crete by Ian Stanton (an ex-member of staff) and the pretty white-flowered dwarf *Lavandula angustifolia* 'Nana Alba'. There is also a dusty pink cultivar, *L. a.* 'Rosea'. We must watch all these woody plants to make sure they are kept fresh and well shaped by timely grooming.

A curious form of sea holly, *Eryngium variifolium*, makes a striking impact among the soft woolly plants. From a basal rosette of heavily veined dark green

leaves rise upright branched stems with leaves and flower heads reduced to long narrow spikes, so silvered they look as if they had been cut out of aluminium. What a superb adaptation to drought!

The variation in thymes always surprises me, both in habit and times of flowering. *Thymus neiceffii*, falling like water over low walls was smothered earlier with mauve-pink flowers over narrow, blue-grey leaves, while *T. serpyllum* 'Pink Chintz' makes loose cushions of soft woolly leaves, buried now beneath lilac-pink flowers. *Thymus* 'Silver Queen', also flowering now, makes attractive small bushes of variegated leaves. Good to cook with, as well as to look at. I will spare you the rest. Another group, too many to describe, are the erodiums, refined relatives of cranesbills, with exquisite silk-petalled flowers in shades of pink and lilac – one is palest sulphur-yellow – and some with the distinctive mark of dark smudges on two of the petals.

The bush-like form of *Nierembergia caerulea* is crowded now with convolvulus-like flowers – white opened-faced bells with blue veins, draining down into blue-stained hearts – while an elegant grass, *Helictotrichon sempervirens* with vivid grey-blue steely stems, adds strength to soft-foliaged groups.

Beyond, outlined in a void against the gravel path, I see the slender form of an American native, *Zigadenus elegans*, just opening on branched spires a succession of small, starry flowers – pale green petals, each with a glistening green patch at the heart, surrounding the ovary. Zigadenus enchants me in this setting, isolated above low mat plants, whereas it might appear insignificant in a general mixed border.

The first season has brought surprises and disappointments. A few things have failed. Odd plants out of a group of the same – *Aurinia saxatilis* (syn. *Alyssum saxatile*), *Crepis incana*, *Anthemis tinctoria* 'E.C. Buxton' – have fainted and failed while others show no sign of distress. There must be a reason. We haven't the answer. I mention them to show that apparently healthy plants can fail inexplicably anywhere, even when given every care. Some plants have grown out of character where easy living has encouraged too much growth. Conditions will vary in the future, both soil and climate. Some over-vigorous plants will be replaced, making room for new introductions.

The beds are not yet mulched with gravel. We have hand-weeded them, using small onion hoes, an almost luxurious job among all this *joie de vivre*. By autumn we shall see, when the final grooming and dying down takes place, where the stone cover will be desirable. Then, in the spring, when there will be fewer germinating weed seeds after the first year, the gravel, about 5cm/2in deep, will be spread.

In conclusion, I repeat what I often say here. If you choose plants adapted by nature to the conditions you have to offer, they will do well, and the garden will give you a sense of peace and fulfilment.

Overall Plan   Island bed No. 1

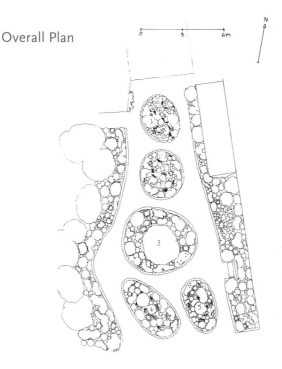

SCREE GARDEN PLANS

The overall plan shows the five scree beds in the setting of surrounding borders, which I have not described in this book since they embrace a somewhat different style of planting – one against a warm west-facing wall, the other a cooler border facing east. The Judas tree is in bed No. 3. The scree beds receive full sun throughout the day.

The detailed plan is of island bed No. 1, where I wished to show off some of our sempervivum collection, built up and cared for by Emily Paston. The raoulias, silver and green, spread rapidly like moss between the compact plants, but will doubtless be checked by frost. To protect them from alternate wetting and freezing, a few patches are sheltered beneath panes of glass. I have hesitated to plant spring delights, such as crocus and chionodoxas, whose lush foliage often smothers small-scale plants, but have chosen species fritillaries and tulips, which seem to be more suitable.

1   *Erodium chrysanthum*
2   *Hebe ochracea* 'James Stirling'
3   *Dianella caerulea*
4   *Thymus* 'Peter Davis'
5   *Euonymus fortunei* 'Golden Pillar'
6   *Thymus* 'Silver Queen'
7   *Yucca filamentosa* 'Variegata'
8   *Thymus vulgaris* 'Golden Pins'
9   *Thymus pulegioides* 'Bertram Anderson'
10   *Erodium cheilanthifolium*
11   *Primula auricula*
12   *Geranium sessiliflorum* subsp. *novae-zelandiae* 'Nigricans'
13   *Sempervivum* 'Hester'
14   *Sedum hispanicum*
15   *Sempervivum* 'Silver Thaw'
16   *Thymus neiceffii*
17   *Sisyrinchium angustifolium*
18   *Fritillaria pyrenaica*
19   *Thymus* 'Hartington Silver'

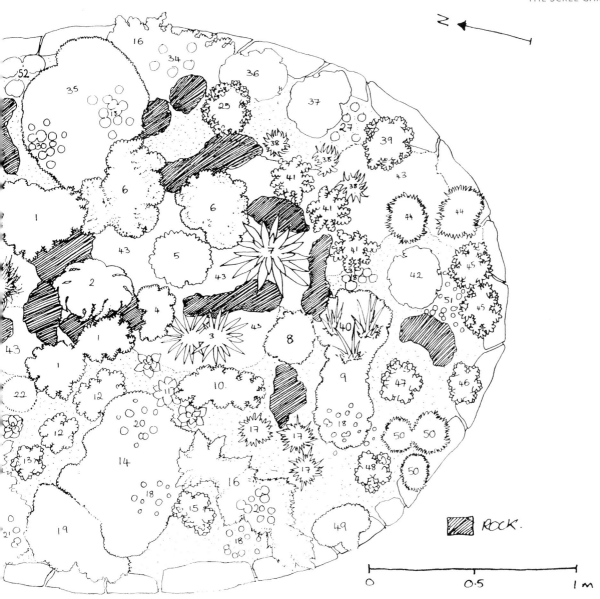

ROCK.

0   0.5   1 m

20 Anemone pavonina
21 Fritillaria tuntasia
22 Armeria juniperifolia 'Bevan's Variety'
23 Sempervivum 'Brock'
24 Dianthus gratianopolitanus 'Tiny Rubies'
25 Sempervivum arachnoideum subsp. tomentosum
26 Raoulia hookeri
27 Tulipa linifolia
28 Androsace sempervivoides
29 Sempervivum tectorum 'Sunset'
30 Fritillaria messanensis

31 Koeleria glauca
32 Sempervivum 'Engle's'
33 Sedum spathulifolium
34 Tulipa linifolia Batalinii Group
35 Sedum hispanicum
36 Anaphaloides bellidioides
37 Antennaria dioica 'Rotes Wunder'
38 Sisyrinchium idahoense
39 Sempervivum octopodes
40 Libertia peregrinans
41 Rhodiola rosea
42 Hebe cupressoides 'Boughton Dome'

43 Raoulia tenuicaulis
44 Leontopodium wilsonii
45 Sempervivum 'Silverine'
46 Sempervivum arachnoideum 'Clärchen'
47 Sempervivum cilosum var. borisii
48 Sempervivum 'Woolcott's Variety'
49 Sedum hispanicum var. minus 'Aureum'
50 Dianthus × arvernensis
51 Tulipa dasystemon
52 Oxalis adenophylla
53 Festuca punctoria

BASIC PLANTS IN THE GRAVEL GARDEN

Tall Verticals

Alcea rugosa
Allium ampeloprasum
Allium hollandicum
Anchusa azurea
Asphodelus albus
Astelia chathamica
Echinops sphaerocephalus
Eremurus cvs
Eremurus stenophyllus
Kniphofia 'Nobilis'
Lilium candidum
Nectaroscordum siculum subsp. bulgaricum
Nepeta nuda subsp. albiflora
Onopordum acanthium
Phlomis tuberosa 'Amazone'
Verbascum bombyciferum
Verbena bonariensis
Yucca gloriosa 'Nobilis'

Medium-sized Plants, Making Vertical Accents

Agapanthus, in variety
Allium atropurpureum
Allium sphaerocephalon
Alstroemeria cvs
Asphodeline liburnica
Asphodeline lutea
Bupleurum falcatum
Eryngium agavifolium
Galtonia candicans and G. viridiflora
Gaura lindheimeri
Kniphofia cvs (medium-sized)
Kniphofia triangularis subsp. triangularis
Linaria purpurea (pink, purple and white)
Lychnis coronaria
Origanum laevigatum 'Hopleys'
Ornithogalum pyrenaicum
Salvia pratensis Haematodes Group
Verbascum chaixii
Verbascum phoeniceum

Short Upright Plants

Anthericum liliago
Gladiolus tristis
Kniphofia 'Little Maid'
Nepeta tuberosa
Salvia nemorosa subsp. tesquicola

Grasses that Create Good Verticals

Calamagrostis × acutiflora 'Karl Foerster'
Elymus magellanicus
Helictotrichon sempervirens
Leymus arenarius
Melica altissima 'Atropurpurea'
Miscanthus sinensis 'Yaku-jima'
Poa labillardierei
Stipa calamagrostis
Stipa gigantea
Stipa tenuissima

Annuals

Eschscholzia californica (cream form)
Nigella damascena 'Miss Jekyll Alba'
Omphalodes linifolia
Papaver commutatum 'Ladybird'
Papaver rhoeas Mother of Pearl Group
Papaver somniferum

Large Shrubs

Abutilon vitifolium 'Album'
Artemisia abrotanum
Atriplex halimus
Berberis × ottawensis
Berberis × stenophylla
Brachyglottis (Dunedin Group) 'Sunshine'
Buddleja alternifolia
Ceanothus thyrsiflorus var. repens
Cistus × cyprius
Cistus × hybridus
Cistus populifolius
Cistus × purpureus

Colutea arborescens

Cotoneaster dammeri

Cotoneaster franchetii

Cotoneaster salicifolius

Euonymus fortune 'Emerald 'n' Gold'

Euphorbia characias subsp. *wulfenii*

Hippocrepis emerus

Lavatera × clementii 'Barnsley'

Lavatera × clementii 'Bredon Springs'

Lavatera × clementii 'Rosea'

Lespedeza bicolor

Ligustrum ovalifolium 'Argenteum'

Ligustrum ovalifolium 'Aureum'

Lonicera nitida 'Baggesen's Gold'

Lonicera pileata

Lonicera × purpusii

Lupinus arboreus

Melianthus major

Phlomis fruticosa

Rhus typhina

Ribes speciosum

Rosa glauca

Rosa spinosissima

Spartium junceum

Tamarix parviflora

Smaller Shrubs

Artemisia in variety

Ballota

Caryopteris × clandonensis cvs

× Halimiocistus wintonensis 'Merrist Wood
   Cream'

Halimium lasianthum

Helianthemum, in variety

Lavandula angustifolia 'Hidcote'

Lavandula angustifolia 'Loddon Pink'

Lavandula angustifolia 'Munstead'

Lavandula angustifolia
   'Twickel Purple'

Lavandula lanata

Lavandula stoechas cvs

Salvia officinalis cvs

Santolina, in variety

Trees

Ailanthus altissima

Amelanchier canadensis

Amelanchier lamarckii

Arbutus unedo

Crataegus persimilis 'Prunifolia'

Eucalyptus dalrympleana

Eucalyptus gunnii

Eucalyptus nicholii

Fraxinus angustifolia 'Raywood'

Genista aetnensis

Juniperus scopulorum 'Skyrocket'

Koelreuteria paniculata

Salix exigua

Spartium junceum

Bulbs

Allium (see page 76)

Anemone blanda

Anemone × fulgens

Anemone pavonina

Arum creticum

Arum italicum subsp. *italicum*
   'Marmoratum'

Camassia leichtlinii subsp. *leichtlinii*

Chionodoxa

Codonopsis grey-wilsonii

Colchicum agrippinum

Colchicum bivonae

Colchicum 'Lilac Wonder'

Colchicum 'Rosy Dawn'

Crocus ligusticus

Crocus nudiflorus

Crocus speciosus

Crocus speciosus 'Albus'

Dierama pulcherrimum

Eremurus stenophyllus

Fritillaria (see page 38)

Galanthus 'James Backhouse'

Galanthus plicatus 'Washfield Warham'

Galtonia candicans

Galtonia princeps

Geranium tuberosum

Gladiolus × colvillii

Gladiolus communis

Gladiolus papilio

Gladiolus tristis

Ipheion uniflorum

Ipheion uniflorum 'Froyle Mill'

Ipheion uniflorum 'Wisley Blue'

Lilium candidum

Nerine bowdenii

Nerine flexuosa Flexuosa Group 'Alba'

Nerine undulata

Ornithogalum montanum

Ornithogalum pyrenaicum

Oxalis adenophylla

Oxalis depressa

Puschkinia scilloides

Scilla bifolia

Scilla siberica

Sternbergia lutea

Triteleia laxa

Tropaeolum polyphyllum

Tropaeolum tuberosum

Tulbaghia

Tulipa dasystemon

Tulipa hageri

Tulipa linifolia

Tulipa orphanidea

Tulipa saxatilis

Zigadenus elegans

**OPPOSITE, FROM TOP
TO BOTTOM** *Nectaroscordum siculum*
subsp. *bulgaricum, Cistus populifolius,
Nigella damascena* 'Miss Jekyll Alba' and
Ribes speciosum.

BASIC PLANTS IN THE SCREE GARDEN

Accent Plants and Verticals

Artemisia alba 'Canescens'
Asphodelus fistulosus
Ceratostigma willmottianum
Dierama 'Puck'
Eryngium bourgatii
Eryngium variifolium
Euonymus fortunei
Hebe ochracea 'James Stirling'
Helichrysum stoechas 'White Barn'
Helicotrichon sempervirens
Hyssopus officinalis 'Roseus'
Lavandula (dwarf forms)
Libertia ixioides
Libertia peregrinans
Linum narbonense
Nepeta tuberosa
Perovskia 'Filigran'
Rhodiola heterodonta
Rhodiola rosea
Salvia × *sylvestris* 'Rose Queen'
Sedum aizoon 'Euphorbioides'
Sedum telephium 'Möhrchen'
Yucca filamentosa 'Variegata'
Zigadenus elegans

Grasses

Festuca glauca 'Elijah Blue'
Helictotrichon sempervirens

Low Creeping Plants

Achillea ageratifolia
Achillea × *lewisii* 'King Edward'
Achillea tomentosa
Anaphaloides bellidioides
Androsace sempervivoides
Antennaria dioica 'Rubra'
Arabis procurrens 'Variegata'
Artemisia glacialis
Artemisia schmidtiana
Artemisia stelleriana

Artemisia stelleriana 'Nana'
Aubrieta 'Silberrand'
Convolvulus sabatius
Cotula lineariloba
Draba aizoides
Eriogonum umbellatum
Frankenia thymifolia
Geranium sessiliflorum subsp. *novae-zelandiae*
    'Nigricans'
Gypsophila repens 'Dubia'
Gypsophila repens 'Rosa Schönheit'
Hypericum empetrifolium subsp. *oliganthum*
Hypericum olympicum f. *minus*
Leontopodium wilsonii
Linum suffruticosum subsp. *salsoloides*
    'Prostratum'
Lithodora diffusa 'Heavenly Blue'
Nepeta nervosa
Onosma alborosea
Origanum amanum
Origanum 'Barbara Tingey'
Origanum 'Kent Beauty'
Origanum rotundifolium
Petrorhagia saxifraga 'Rosette'
Phlox douglasii 'Boothman's Variety'
Phlox 'Kelly's Eye'
Phlox × *procumbens* 'Variegata'
Phlox subulata 'Lilacina'
Phlox subulata 'McDaniel's Cushion'
Phlox subulata 'Temiskaming'
Phlox subulata 'White Delight'
Raoulia
Salvia lavandulifolia subsp. *blancoana*
Saponaria 'Bressingham'
Saponaria ocymoides
Saponaria ocymoides 'Alba'
Saponaria × *olivana*
Sedum hispanicum
Sedum hispanicum var. *minus* 'Aureum'
Sedum spathulifolium
Sedum spathulifolium 'Cape Blanco'

Sedum spurium 'Atropurpureum'
Sedum spurium 'Tricolor'
Sedum telephium subsp. *ruprechtii*
Silene uniflora 'Druett's Variegated'
Silene uniflora 'Robin Whitebreast'
Tanacetum densum subsp. *amani*
Thymus Coccineus Group
Thymus doerfleri
Thymus 'Golden King'
Thymus longicaulis
Thymus neiceffii
Thymus pugelioides 'Bertram Anderson'
Thymus serpyllum var. *albus*
Thymus serpyllum 'Minor'
Thymus serpyllum 'Pink Chintz'
Thymus 'Silver Queen'
Veronica cinerea
Veronica umbrosa 'Georgia Blue'
Zauschneria californica 'Ed Carman'
Zauschneria californica 'Olbrich Silver'

Bulbs and Bulbous Plants

Allium flavum
Allium karataviense
Allium obliquum
Anemone multifida
Anemone pavonina
Asphodelus fistulosus
Freesia laxa
Fritillaria michailovskyi
Fritillaria pyrenaica
Fritillaria tuntasia
Nerine bowdenii
Nerine filifolia
Nerine undulata
Tulbaghia violacea
Zigadenus elegans

Small to Medium Clump-forming Plants

Agapanthus 'Windlebrooke'
Alyssum spinosum 'Roseum'
Antirrhinum hispanicum subsp. *hispanicum*
Armeria alliacea
Armeria juniperifolia 'Bevan's Variety'
Armeria maritima 'Corsica'
Armeria maritima 'Düsseldorfer Stolz'
   (Düsseldorf Pride)
Artemisia alba 'Canescens'
Aurinia saxatilis 'Citrina'
Aurinia saxatilis 'Dudley Nevill Variegated'
Bergenia stracheyi
Convolvulus cneorum
Crepis aurea
Crepis incana
Dianthus, in variety
Diascia barberae 'Blackthorn Apricot'
Diascia barberae 'Ruby Field'
Diascia Coral Belle
Diascia rigescens
Diascia vigilis
Diplarrena Helen Dillon's form
Diplarrena moraea
Erodium × *kolbianum* 'Natasha'
Helichrysum 'Schwefellicht' (Sulphur Light)
Hypericum coris
Hypericum olympicum
Hypericum olympicum f. *uniflorum*
   'Citrinum'
Iberis sempervirens 'Weisser Zwerg'
Limonium bellidifolium
Limonium binervosum
Linum flavum 'Compactum'
Lychnis flos-jovis
Nierembergia caerulea
Origanum amanum
Origanum 'Barbara Tingey'
Origanum 'Kent Beauty'
Osteospermum jucundum

Osteospermum 'Lady Leitrim'
Othonna cheirifolia
Primula auricula (mixed forms)
Pulsatilla vulgaris
Pulsatilla vulgaris 'Alba'
Rhodanthemum hosmariense
Rhodiola heterodonta
Rhodiola rosea
Santolina chamaecyparissus 'Nana'
Sedum aizoon 'Euphorbioides'
Sempervivum
Serratula seoanei
Sideritis syriaca
Silene keiskei
Silene schafta
Sisyrinchium angustifolium
Sisyrinchium 'Californian Skies'
Sisyrinchium idahoense
Veronica austriaca subsp. *teucrium*
   'Kapitän'

BIBLIOGRAPHY

Grounds, Roger, *Ornamental Grasses* (1981), Van Nostrand, London and New York

Hillier's Manual of Trees and Shrubs (5th edn, 1981), David & Charles, Newton Abbot, Devon

Ingwersen, Will, *Ingwersen's Manual of Alpine Plants* (1979), W. Ingwersen and Dunnsprint, Eastbourne

King, Michael and Oudolf, Piet, *Gardening With Grasses* (1998), Frances Lincoln, London, and Timber, Portland

Lloyd, Christopher, *Clematis* (1989), Viking, London

Lloyd, Christopher, *The Well-Tempered Garden* (1989), Viking, London

Mathew, Brian, *Dwarf Bulbs* (1973), Batsford, London

Phillips, Roger and Rix, Martyn, *The Bulb Book* (1981), Pan, London

RHS Plant Finder 1999–2000 (1999), Dorling Kindersley, London

Stuart Thomas, Graham, *Perennial Garden Plants: Or the Modern Florilegium: A Concise Acount of Herbaceous Plants, Including Bulbs, for General Garden Use* (1982), J. M. Dent and Sons, London

Stuart Thomas, Graham, *Shrub Roses of Today* (1983), J. M. Dent and Sons Ltd, London

INDEX

AUTHOR'S ACKNOWLEDGMENTS

This book is the result of a great team effort.

My husband, Andrew, was a key member of this team. His life-long studies have inspired and guided my gardening life. He shared my delight in the development of the garden, then died peacefully in a room overlooking the Scree Garden on 20 September 1999, aged ninety years.

Erica Hunningher took the notes I had written over a period of several years, which resembled a jumbled bag of patchwork pieces. Erica painstakingly sorted them out and with a clear vision stitched them together to form a seamless quilt.

Tricia Brett, my secretary and PA, scoured the word processor, typing and retyping with great patience to make sure no scraps were mislaid.

Both she and David Ward, my nursery manager, have helped with proofreading. Together with the rest of my staff they have ensured the smooth running of the nursery and garden, all supporting me at a time when I needed them most.

Another talented member of my team, Chrissy McDonald, has drawn and painted the endpapers of this book. She also drew the plans of two individual beds, on pages 28–9 and 180–1. The original overall plan of the Gravel Garden was drawn to scale by Jill Billington, with a party of students from Capel Manor. The overall plan of the Scree Garden was drawn to scale by Denis Lloyd, a tutor at Writtle College, assisted by my student Yuko Tanabe.

Sarah Pickering has used both restraint and sensitivity in designing this book. She has created a balance, both for ease of reading and for losing oneself in the pictures.

To Tony Lord (editor of *RHS Plant Finder*) I am indebted for bringing my plant names up to date.

I would also like to thank Brian Cook of Hythe Offset, Colchester, for printing the garden plan used on the endpapers.

Finally, but by no means least, I am indebted to Steve Wooster for his skill in capturing fleeting moments on film. Too often I have despaired when I thought the best moments were lost, the camera not here or the weather tossing everything awry. But when he appears, Steve finds pictures I had not dreamt of. It is the mind and eye behind the camera which creates its own vision of the garden.

PHOTOGRAPHER'S NOTE

I have been photographing the gardens for some time now, after working as the designer, many years ago, on *Beth Chatto's Green Tapestry*. This was really the first book for which I started to take pictures, in an effort to get me away from my desk occasionally. Style and quality of photography (and film) have moved on greatly since those days.

It was three or four years ago that I made a visit to the gardens after a longish break, and I was surprised to see the transformation to what used to be a car park. I think it was late June and there were plants and flowers everywhere, as though they had been there always. I knew from that day that there would be more than enough material there to fill a book. That must speak volumes for the quality and intensity of the planting, as the area that the Gravel Garden covers is relatively small.

I was very fortunate to have more than a year to record the garden. My natural liking is for big, bold, architectural-style planting, so the first pictures (shot on medium format) had this feeling. As the book progressed the pictures tended to become more 'intimate' and small-scale. Some of the last pictures were extreme close-ups of water droplets on foliage and the beards of iris flowers. I am indebted to Beth for pointing out many little gems to me – plants that I may have trodden on in the past to get at the big stuff!

Because of the quality of the planting there was always something to photograph at any time of year. The weather wasn't always equally accommodating, as there was often a wind blowing, which is so frustrating for garden photographers. Early mornings were sometimes quite still, but on a sunny day in midsummer it could often be too bright by about 8.00 a.m. – the sun would suddenly flood in over the leylandii hedge, which shelters the Gravel Garden from the north-east. Another advantage of early morning photography is that there aren't any people about. However, I naturally prefer taking pictures in the evening for the softness of the light, although morning shots are essential for dew in summer and frost in winter. My thanks go to Tricia for weather forecasting, to David and to the good-natured and cheerful members of staff who provided assistance when needed.

PUBLISHER'S ACKNOWLEDGMENTS

Art Editor Sarah Pickering
Text Editor Anne Askwith
Paperback text editor Joanna Chisholm
Editorial assistance Tom Windross

Production Hazel Kirkman
Index compiled by Marie Lorimer
Index revised by Michele Clarke
Horticultural Consultant Tony Lord

Editorial Director Kathryn Cave
Art Director Caroline Hillier
Head of Pictures Anne Fraser